DISCARDED

C0-BXA-409

Lecture Notes in Statistics 110

Edited by P. Bickel, P. Diggle, S. Fienberg, K. Krickeberg,
I. Olkin, N. Wermuth, S. Zeger

DISCARDED

Springer
New York
Berlin
Heidelberg
Barcelona
Budapest
Hong Kong
London
Milan
Paris
Santa Clara
Singapore
Tokyo

D. Bosq

Nonparametric Statistics for Stochastic Processes

Estimation and Prediction

Springer

D. Bosq
Université Pierre et Marie Curie
Institut de Statistique
4 Place Jussieu
75252 Paris cedex 05
France

Library of Congress Cataloging-in-Publication Data Available
Printed on acid-free paper.

© 1996 Springer-Verlag New York, Inc.
All rights reserved. This work may not be translated or copied in whole or in part without the written permission of the publisher (Springer-Verlag New York, Inc., 175 Fifth Avenue, New York, NY 10010, USA), except for brief excerpts in connection with reviews or scholarly analysis. Use in connection with any form of information storage and retrieval, electronic adaptation, computer software, or by similar or dissimilar methodology now known or hereafter developed is forbidden. The use of general descriptive names, trade names, trademarks, etc., in this publication, even if the former are not especially identified, is not to be taken as a sign that such names, as understood by the Trade Marks and Merchandise Marks Act, may accordingly be used freely by anyone.

Camera ready copy provided by the author.
Printed and bound by Braun-Brumfield, Ann Arbor, MI.
Printed in the United States of America.

9 8 7 6 5 4 3 2 1

ISBN 0-387-94713-2 Springer-Verlag New York Berlin Heidelberg   SPIN 10532130

To MARIE, CAMILLE and ROMANE.

Preface

Recently new developments have taken place in the theory of nonparametric statistics for stochastic processes. Optimal asymptotic results have been obtained and special behaviour of estimators and predictors in continuous time has been pointed out.

This book is devoted to these questions. It also gives some indications about implementation of nonparametric methods and comparaison with parametric ones, including numerical results. Many of the results presented here are new and have not yet been published, expecially those in Chapters IV and V.

I am grateful to W. Härdle, Y. Kutoyants, F. Merlevede and G. Oppenheim who made important remarks that helped much to improve the text.

I am greatly indebted to B. Heliot for her careful reading of the manuscript which allowed to ameliorate my english. I also express my gratitude to D. Blanke, L. Cotto and P. Piacentini who read portions of the manuscript and made some useful suggestions.

I also thank M. Gilchrist and J. Kimmel for their encouragements.

My acknowledgment also goes to M. Carbon, M. Delecroix, B. Milcamps and J.M. Poggi who authorized me to reproduce their numerical results.

My greatest debt is to D. Tilly who prepared the typescript with care and efficiency.

Notations

A^c, $A \cup B$, $A \cap B$ complement of A, union of A and B, intersection of A and B.

$\overset{\circ}{A}$, $\overline{A}$ interior of A, closure of A.

$(\Omega, \mathcal{A}, P)$ Probability space : Ω non empty set, $\mathcal{A}$ σ-Algebra of subsets of Ω, P Probability measure on $\mathcal{A}$.

$\mathcal{B}_{\mathbb{R}^d}$ σ-Algebra of Borel sets on $\mathbb{R}^d$.

$\sigma(X_i, \ i \in I)$ σ-Algebra generated by the random variables X_i, $i \in I$.

i.i.d. r.v.'s independent and identically distributed random variables.

EX, VX, P_X, f_X expectation, variance, distribution, density (of X).

$E(X \mid \mathcal{B})$, $E(X \mid X_i, i \in I)$, $V(X \mid \mathcal{B})$, $V(X \mid X_i, i \in I)$ conditional expectation, conditional variance (of X), with respect to $\mathcal{B}$ or to $\sigma(X_i, i \in I)$.

$\mathrm{Cov}(X, Y)$, $\mathrm{Corr}(X, Y)$ covariance, correlation coefficient (of X and Y).

$\delta_{(a)}$, $\mathcal{B}(n, p)$, $\mathcal{N}(m, \sigma^2)$, λ^d Dirac measure, Binomial distribution, normal distribution, Lebesgue measure over $\mathbb{R}^d$.

$(X_t, \ t \in I)$ or (X_t) stochastic process.

$L^p(E, \mathcal{B}, \mu)$ (or $L^p(E)$, or $L^p(\mathcal{B})$, or $L^p(\mu)$) space of (classes) of real $\mathcal{B} - \mathcal{B}_{\mathbb{R}}$ measurable functions f such that
$$\parallel f \parallel_p = \left(\int_E |f|^p d\mu \right)^{1/p} < +\infty \quad (1 \leq p < +\infty),$$
$$\parallel f \parallel_\infty = \inf\{a : \mu\{f > a\} = 0\} < +\infty \quad (p = +\infty).$$

$\displaystyle\int_{\mathbb{R}^d} f(x)dx$ integral of f with respect to Lebesgue measure on $\mathbb{R}^d$.

$\mathbf{1}_A$ indicator of A : $\mathbf{1}_A(x) = 1$, $x \in A$; $= 0$, $x \notin A$.

$\mathrm{Log}_k x$ defined recursively by $\mathrm{Log}_k(x) = \mathrm{Log}(\mathrm{Log}_{k-1} x)$ if $\mathrm{Log}_{k-1} x \geq e$, $\mathrm{Log}_k(x) = 1$ if $\mathrm{Log}_{k-1} x < e$; $k \geq 2$.

$[x]$ integer part of x.

$f \otimes g$ defined by $(f \otimes g)(x, y) = f(x)g(y)$.

$u_n \sim v_n \quad \dfrac{u_n}{v_n} \longrightarrow 1.$

$u_n \simeq v_n$ or $u_n \cong v_n$. There exist constants c_1 and c_2 such that $0 < c_1 v_n < u_n < c_2 v_n$ for n large enough.

$u_n = o(v_n)$ $\dfrac{u_n}{v_n} \longrightarrow 0$.

$u_n = O(v_n)$ $u_n \leq c v_n$ for some $c > 0$.

$\overset{w}{\longrightarrow}$ weak convergence.

$\overset{p}{\longrightarrow}$ convergence in probability.

$\overset{a.s.}{\longrightarrow}$ almost sure convergence.

$\overset{q.m.}{\longrightarrow}$ convergence of quadratic mean.

■ end of a proof.

$\sharp E$ cardinal of E.

Synopsis

S.1 The object of the study

Classically time series analysis has two purposes.One of these is to construct a model which fits the data and then to estimate the model's parameters. The second object is to use the identified model for prediction.

The popular so-called BOX-JENKINS approach gives a complete solution of the above mentioned problems through construction, identification and forecasting of an ARMA process or more generally a SARIMA process (cf. [B-J], [G-M], [B-D]).

Unfortunately the underlying assumption of linearity which supports the B-J's theory is rather strong and, therefore, inadequate in many practical situations.

That inadequacy appears in the forecasts, especially if the horizon is large. Consideration of nonlinear parametric models, like bilinear or ARCH processes, does not seem to give a notable improvement of the forecasts.

On the contrary a suitable nonparametric predictor supplies rather precise forecasts even if the underlying model is truly linear and if the horizon is remote. This fact explains the expansion of nonparametric methods in time series analysis during the last decade. Note however that parametric and nonparametric methods are complementary since a parametric model tries to explain the mechanism which generates the data.

It is important to mention that the duality highlighted at the beginning of the current section is not conspicuous in a nonparametric context because the underlying model only appears through regularity conditions whereas estimating and forecasting are basic.

Figure 1 gives an example of comparison between nonparametric and parametric forecasts. Other numerical comparisons appear in the appendix.

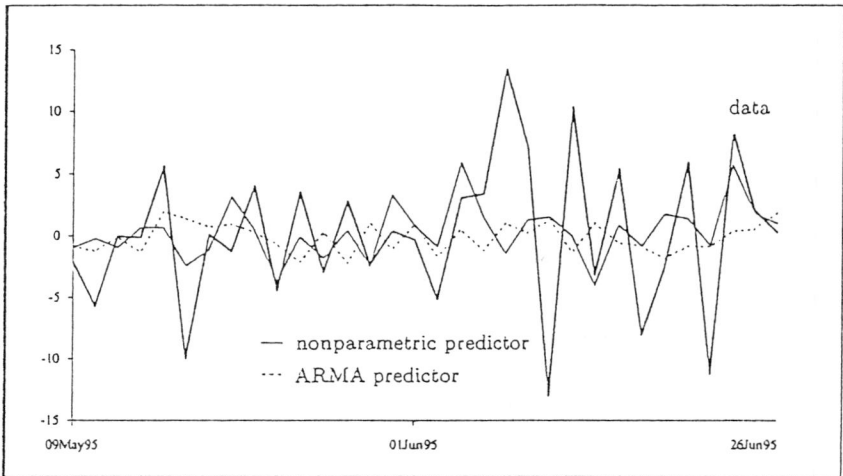

Forecasting of french ten years yields
The nonparametric predictor gives better indications
about signs of variation
Figure 1

In this book we present optimal asymptotic results on density and regression nonparametric estimation with applications to prediction, as well in discrete time as in continuous time.

We also try to explain why nonparametric forecasts are (in general) more accurate than parametric ones. Finally we make suggestions for the implementation of functional estimators and predictors.

Note that we do not pretend to provide an encyclopaedic treatment of nonparametric statistics for stochastic processes. Actually our work focuses on prediction by the kernel method.

Now the rest of the synopsis is organized as follows. In S.2 we construct the kernel density estimator. The kernel regression estimator and the associated predictor are considered in S.3. The mathematical tools defined in Chapter 1 are described in S.4. S.5 deals with the asymptotic behaviour of density estimators (cf. Chapters 2 and 4). S.6 is devoted to the convergence of regression estimators and predictors (cf. Chapters 3 and 4). Finally S.7 discusses sampling, and practical considerations (cf. Chapter 3 and appendix).

S.2 The kernel density estimator

We now describe the popular kernel density estimator. For the sake of simplicity we first suppose that the data $X_1, \ldots, X_n$ come from a sequence of real independent random variables with a common density f belonging to some family $\mathcal{F}$.

If $\mathcal{F}$ is large (for example if $\mathcal{F}$ contains the continuous densities) it is well known that no unbiased estimator of f can exist (see [RO1]) and that the supremum of the likelihood is infinite.

Then, a primary density estimator should be the **histogram** defined as

$$\widehat{f}_n(x) = \frac{\nu_{nj}}{n(a_{n,j} - a_{n,j-1})} \ , \ x \in I_{nj} \ , \ j \in \mathbb{Z},$$

where $I_{nj} = [a_{n,j-1}, a_{n,j}[$ and $(a_{nj}, \ j \in \mathbb{Z})$ is a strictly increasing sequence such that $|a_{jn}| \to \infty$ as $|j| \to \infty$, and where $\nu_{nj} = \sum_{i=1}^{n} \mathbf{1}_{[a_{n,j-1}, a_{n,j}[}(X_i)$.

If f is continuous over I_{nj} and if $a_{n,j} - a_{n,j-1}$ is small, then $\widehat{f}_n(x)$ is close to $f(x)$ for each x in I_{nj}. However this estimator does not utilize all the information about $f(x)$ contained in data since observations which fall barely outside I_{nj} do not appear in $\widehat{f}_n(x)$. This drawback is particularly obvious if $x = a_{n,j-1}$.

A remedy should be the construction of an **adaptable histogram** defined as

$$f_n^*(x) = \frac{\nu_n(x)}{nh_n} \ , \ x \in \mathbb{R},$$

where $\nu_n(x) = \sum_{i=1}^{n} \mathbf{1}_{[x - \frac{h_n}{2}, x + \frac{h_n}{2}]}(X_i)$ and where h_n is a given positive number.

Note that f_n^* may be written under the form

$$f_n^*(x) = \frac{1}{nh_n} \sum_{i=1}^{n} K_0 \left(\frac{x - X_i}{h_n} \right) \ , \ x \in \mathbb{R},$$

where $K_0 = \mathbf{1}_{[-\frac{1}{2}, +\frac{1}{2}]}$ is the so-called **naive kernel**.

The accuracy of f_n^* depends heavily on the choice of the "bandwidth" h_n. This choice must conciliate two contradictory requirements : the smallness of $\left[x - \frac{h_n}{2}, \ x + \frac{h_n}{2} \right]$ and a large number of observations falling in this interval.

Since $E\nu_n(x) \simeq nh_n f(x)$ (provided h_n be small and $f(x) > 0$) we obtain the conditions :

(C_1) $h_n \to 0$, $nh_n \to +\infty$ as $n \to +\infty$.

If the X_i's are $\mathbb{R}^d$-valued, f_n^* is defined as

$$f_n^*(x) = \frac{1}{nh_n^d} \sum_{i=1}^{n} \mathbf{1}_{[-\frac{1}{2}, +\frac{1}{2}]^d} \left(\frac{x - X_i}{h_n} \right) \ , \ x \in \mathbb{R}^d$$

and (C_1) becomes

(C_2) $h_n \to 0$, $nh_n^d \to +\infty$ as $n \to +\infty$.

Now in order to obtain smoother estimations, one can use other **kernels** (a kernel on $\mathbb{R}^d$ is a bounded symmetric d-dimensional density such that $\| u \|^d K(u) \to 0$ as $\| u \| \to \infty$ and $\int \| u \|^2 K(u) du < \infty$). Let K be a kernel, the associated kernel estimator is

$$f_n(x) = \frac{1}{nh_n^d} \sum_{i=1}^{n} K \left(\frac{x - X_i}{h_n} \right) \ , \ x \in \mathbb{R}^d$$

For example if $d = 1$ and if $K(u) = \frac{1}{\sqrt{2\pi}} e^{-\frac{u^2}{2}}$, $u \in \mathbb{R}$ then

$$f_n(x) = \frac{1}{n} \sum_{i=1}^{n} \frac{1}{h_n \sqrt{2\pi}} e^{-\frac{1}{2} \left(\frac{x - X_i}{h_n} \right)^2} \ , \ x \in \mathbb{R}$$

which is a mixture of Gaussian densities with respective means X_i and variance h_n^2.

We now consider the case where the data are realizations of a stochastic process (X_t). In that case the Kolmogorov extension theorem states that the distribution ν of a stochastic process is completely specified by its finite-dimensional distributions (cf. [A.G]). Thus the general problem of estimating ν reduces to the estimation of these. To this aim, it is convenient to estimate the associated densities if they do exist.

If (X_t) is a discrete time process one may use f_n as well. Finally if $(X_t, t \in \mathbb{R})$ denotes a d-dimensional continuous time process observed over the time interval $[0, T]$, the kernel density estimator is defined by setting

$$f_T(x) = \frac{1}{Th_T^d} \int_0^T K \left(\frac{x - X_t}{h_T} \right) dt \ , \ x \in \mathbb{R}^d$$

where h_T is a given positive number.

For other methods of density estimation we refer to [PR]. The method of kernels has several advantages : it is natural, easy to compute and robust; moreover it reaches optimal rates as we shall see in Chapter 4.

S.3 The kernel regression estimator and the induced predictor

In the context of regression estimation the analogue of the histogram is the so-called **regressogram** : let us consider i.i.d. bidimensional random variables $(X_1, Y_1), \ldots, (X_n, Y_n)$ such that a specified version r of the nonlinear regression of Y_i on X_i does exist :

$$r(x) = E(Y_i \mid X_i = x) \quad , \quad x \in \mathbb{R} .$$

The regressogram has the following form

$$\widehat{r}_n(x) = \frac{\displaystyle\sum_{i=1}^n Y_i \mathbf{1}_{I_{nj}}(X_i)}{\displaystyle\sum_{i=1}^n \mathbf{1}_{I_{nj}}(X_i)} \quad , \ x \in I_{nj} \ , \ j \in \mathbb{Z}$$

where $(I_{nj}, j \in \mathbb{Z})$ is defined in S.2. Note that $\widehat{r}_n$ is defined only if $\sum_{i=1}^n \mathbf{1}_{I_{nj}}(X_i) > 0.$

$\widehat{r}_n(x)$ may be interpreted as an estimator of $E(Y_i \mid X_i \in I_{nj})$. It clearly suffers from the same drawback as the histogram and the remedy is similar. This brings us to define the kernel regression estimator (cf. [NA], [WA]) as

$$r_n(x) = \frac{\displaystyle\sum_{i=1}^n Y_i K\left(\dfrac{x - X_i}{h_n}\right)}{\displaystyle\sum_{i=1}^n K\left(\dfrac{x - X_i}{h_n}\right)} \quad , \ x \in \mathbb{R}$$

where (K, h_n) is defined in S.2. The definition remains valid if (X_i, Y_i) is $\mathbb{R}^d \times \mathbb{R}$-valued and if data are dependent.

Now, if (X_t, Y_t), $t \in \mathbb{R}$, is a $\mathbb{R}^d \times \mathbb{R}$-valued continuous time process observed over the time interval $[0, T]$, the kernel regression estimator is defined as

$$r_T(x) = \frac{\displaystyle\int_0^T Y_t K\left(\frac{x - X_t}{h_T}\right) dt}{\displaystyle\int_0^T K\left(\frac{x - X_t}{h_T}\right) dt} \quad , \ x \in \mathbb{R}^d .$$

Let us now turn to **prediction**. For the sake of simplicity we consider a real square integrable Markov process $(\xi_t, \ t \in \mathbb{Z})$ and the data $\xi_1, \ldots, \xi_n$. The problem is to construct a **statistical predictor** of ξ_{n+H} where H is a strictly positive integer.

Such a predictor, say $\widehat{\xi}_{n+H}$, is an approximation of $r(\xi_n) = E(\xi_{n+H} \mid \xi_n)$ based on $\xi_1, \ldots, \xi_n$; and the statistical error of prediction is defined as $E\left(\widehat{\xi}_{n+H} - r(\xi_n)\right)^2$.

Note that the total error of prediction is

$$E(\widehat{\xi}_{n+H} - \xi_{n+H})^2 = E(\widehat{\xi}_{n+H} - r(\xi_n))^2 + E(r(\xi_n) - \xi_{n+H})^2$$

where the last term is structural and consequently cannot be controlled by the statistician.

Now let us consider the bidimensional process $(X_t, Y_t) = (\xi_t, \xi_{t+H})$, $t \in \mathbb{Z}$ and the associated regression estimator r_{n-H} defined above. It induced a natural **nonparametric predictor** via the formula

$$\widehat{\xi}_{n+H} = \frac{\displaystyle\sum_{i=1}^{n-H} \xi_{i+H} K\left(\frac{\xi_n - \xi_i}{h_n}\right)}{\displaystyle\sum_{i=1}^{n-H} K\left(\frac{\xi_n - \xi_i}{h_n}\right)} .$$

Similarly if $(\xi_t, \ t \in \mathbb{R})$ is a real square integrable Markov process observed over $[0, T]$, the nonparametric predictor of ξ_{T+H} associated with r_{T-H} is given by

$$\widehat{\xi}_{T+H} = \frac{\displaystyle\int_0^{T-H} \xi_{t+H} K\left(\frac{\xi_T - \xi_t}{h_T}\right) dt}{\displaystyle\int_0^{T-H} K\left(\frac{\xi_T - \xi_t}{h_T}\right) dt} .$$

The final purpose of this book is the asymptotic study of $\widehat{\xi}_{n+H}$, $\widehat{\xi}_{T+H}$ and of some more general predictors.

S.4 Mixing processes

In order to obtain rates of convergence for functional estimators it is necessary to have measures of dependence between the observed variables at one's disposal.

Some measures of that type are introduced in Chapter 1. The most important should be the strong mixing coefficient (cf. [R0]). For the sake of clarity, let us introduce it in a stationary context.

Let $X = (X_t, \ t \in \mathbb{Z})$ be a strictly stationary[1] process, its strong mixing coefficient of order k is defined as

$$\alpha(k) = \sup_{\substack{B \in \sigma(X_s, \ s \leq t) \\ C \in \sigma(X_s, \ s \geq t + k)}} |P(B \cap C) - P(B)\, P(C)| \ , \quad k \geq 1$$

X is said to be **strongly mixing** (or **α-mixing**) if $\lim\limits_{k \to \infty} \alpha(k) = 0$. This condition specifies a form of asymptotic independence of the past and future of X.

Classical ARMA processes are strongly mixing with coefficients which decrease to zero at an exponential rate.

Now if X is strongly mixing it is possible to derive some useful **covariance inequalities**. An example is the following : if $Y \in L^\infty(\sigma(X_s, \ s \leq t))$ and $Z \in L^\infty(\sigma(X_s, \ s \geq t + k))$ then

$$|Cov(Y,Z)| \leq 4 \parallel Y \parallel_\infty \parallel Z \parallel_\infty \alpha(k).$$

Among these inequalities the sharper is due to RIO (cf. [RI] 1993). RIO's inequality is optimal in some sense (see (1.9) and Theorem 1.1).

Other important outcomes of the strong mixing condition are **large deviation inequalities**. An accurate lemma of BRADLEY (Lemma 1.2) gives the "cost" of the replacement of dependent random variables by associated independent ones. Using this result and exponential type inequalities for independent variables it is thus possible to establish large deviation inequalities for strongly mixing processes.

---

[1]Let us recall that X is said to be **(strictly) stationary** if, for any integer k and any $t_1, \ldots, t_k$, s in $\mathbb{Z}$ one has $P_{\left(X_{t_1+s}, \ldots, X_{t_k+s}\right)} = P_{\left(X_{t_1}, \ldots, X_{t_k}\right)}$. For such a process $\alpha(k)$ does not depend on t.

As an example we give the following (cf. Theorem 1.3). Let $(X_t, \ t \in \mathbb{Z})$ be a zero-mean real-valued strictly stationary bounded process. Then for each integer $q \in \left[1, \dfrac{n}{2}\right]$ and each $\varepsilon > 0$

$$P\left(\left|\sum_{t=1}^{n} X_t\right| > n\varepsilon\right) \leq 4\exp\left(-\frac{\varepsilon^2}{8v^2(q)}q\right)$$
$$+ \ 22\left(1 + \frac{4\|X_0\|_\infty}{\varepsilon}\right)^{1/2} q\alpha\left(\left[\frac{n}{2q}\right]\right)$$

where

$$v^2(q) = \frac{8q^2}{n}V\left(\sum_{t=1}^{\left[\frac{n}{2q}+1\right]} X_t\right) + \frac{\varepsilon\|X_0\|_\infty}{2}$$

and $\alpha\left(\left[\dfrac{n}{2q}\right]\right)$ is the strong mixing coefficient of order $\left[\dfrac{n}{2q}\right]$.

This inequality allows to derive limit theorems for strongly mixing processes (cf. Theorems 1.5, 1.6, 1.7).

S.5 Density Estimation

S.5.1 Discrete case

Chapter 2 deals with density estimation for discrete time processes. The main problem is to achieve the **optimal rates**, that is the same rates as in the i.i.d. case.

First it can be shown that, under some regularity assumptions, if f is twice differentiable and if $(X_t, \ t \in \mathbb{Z})$ satisfies a mild mixing condition then, for a suitable choice of (h_n),

$$n^{4/(d+4)}E(f_n(x) - f(x))^2 \longrightarrow c$$

where c is explicit (Theorem 2.1). Thus the optimal rate in quadratic mean is achieved. The proof uses the covariance inequality stated in S.4.

Concerning uniform convergence it may be proved that for each $k \geq 1$ we have

$$\sup_{x \in \mathbb{R}^d} |f_n(x) - f(x)| = o\left(\text{Log}_k n\left(\frac{\text{Log}\,n}{n}\right)^{2/(d+4)}\right) \qquad \text{a.s.}$$

(cf. Corollary 2.2). This result is (almost) optimal since the uniform rate of convergence in the i.i.d. case is $O\left(\left(\dfrac{\mathrm{Log}\,n}{n}\right)^{2/(d+4)}\right)$.

Here the main assumption is that (X_t) is strongly mixing with $\alpha(k) \le a\rho^k$ $(a > 0,\ 0 < \rho < 1)$, and the proof uses the large deviation inequality presented in S.4.

We also establish the following weak convergence result (Theorem 2.3) :

$$\left(nh_n^d\right)^{1/2}\left(\frac{f_n(x_i) - f(x_i)}{(f_n(x_i))^{1/2}\,\|\,K\,\|_2},\ 1 \le i \le m\right) \xrightarrow{w} N^{(m)}$$

where $N^{(m)}$ has the m-dimensional standard normal distribution. Note that the precise form of this result allows to use it for constructing tests and confidence sets for the density.

Here $\alpha(k) = O(k^{-2})$, and the proof utilizes the BRADLEY lemma quoted in S.4.

The end of Chapter 2 is devoted to the asymptotic behaviour of f_n in some unusual situations : chaotic data, singular distribution, processes with errors in variables.

S.5.2   Continuous case

The problem of estimating density for continuous time processes is investigated in Chapter 4.

The search for optimal rates is performed in a more general setting than in discrete time, here f is supposed to be k times differentiable with k^{th} partial derivatives satisfying a Lipschitz condition of order λ $(0 < \lambda \le 1)$. Thus the number $r = k + \lambda$ characterizes the regularity of f. In that case it is interesting to choose K in a special class of kernels (cf. Section 4.1).

Then it can be shown that under mild regularity conditions

$$\lim_{T\to\infty}\ \sup_{X\in\mathcal{X}_1}\ \sup_{x\in\mathbb{R}^d} T^{2r/(2r+d)}E_X(f_T(x) - f(x))^2 < +\infty$$

where $\mathcal{X}_1$ denotes a suitable family of continuous time processes (Corollary 4.2). Furthermore the rate $T^{-2r/(2r+d)}$ is **minimax** (Theorem 4.3).

Now this rate is achieved if the observed sample paths are slowly varying, otherwise the rate is more accurate.

The phenomenon was first pointed out by CASTELLANA and LEADBET-TER in 1986 (cf. [C-L]). The following is an extension of their result : if the density $f_{(X_s,X_t)}$ exists for all (s,t), $s \neq t$ and if for some $p \in [1, +\infty]$ we have

$$(C_p) \quad \limsup_{T \to \infty} \frac{1}{T} \int_{]0,T]^2} \| f_{(X_s,X_t)} - f \otimes f \|_p \, dsdt < \infty$$

then

$$\sup_{x \in \mathbb{R}^d} E(f_T(x) - f(x))^2 = O\left(T^{-pr/(pr+d)}\right)$$

(Theorem 4.6); in particular if (C_∞) holds then

$$\sup_{x \in \mathbb{R}^d} E(f_T(x) - f(x))^2 = O\left(\frac{1}{T}\right).$$

From now on $\frac{1}{T}$ will be called "superoptimal rate" or "parametric rate".

Condition (C_p) first measures the asymptotic independence between X_s and X_t when $|t - s|$ is large, second, and above all, the local behaviour of $f_{(X_s,X_t)}$ when $|t - s|$ is small.

If p is large enough $(p > 2)$ the local irregularity of the sample paths furnishes additional information. This explains the improvement of the so called "optimal rate".

The situation is especially simple in the Gaussian case : if (X_t) is a real stationary Gaussian process, regular enough and if K is a strictly positive kernel, then Corollary 4.4 entails the following alternative :

- If $\int_0^\varepsilon \left(E|X_u - X_0)^2\right)^{-1/2} du < \infty$ then $E(f_T - f)^2 = O\left(\frac{1}{T}\right)$

- If $\int_0^\varepsilon \left(E|X_u - X_0|^2\right)^{-1/2} du = \infty$ then $TE(f_T - f)^2 \to \infty$.

In particular if (X_t) has differentiable sample paths the superoptimal rate is **not** achieved.

Now the same phenomenon appears in the study of uniform convergence : using a special Borel-Cantelli lemma for continuous time processes (cf. Lemma 4.2) one can obtain an optimal rate under mild conditions, but also a superoptimal rate under stronger conditions. In fact it can be proved that

$$\sup_{x \in \mathbb{R}^d} |f_T(x) - f(x)| = o\left(\left(\frac{\text{Log}T}{T}\right)^{1/2} \text{Log}_k T\right) \quad \text{a.s.}, \quad k \geq 1 .$$

S.6 Regression estimation and prediction

S.6.1 Regression estimation

Contrary to the density, the regression cannot be consistently estimated uniformly over the whole space.

This because the magnitude of $r(x)$ for $\| x \|$ large is unpredictable. However it is possible to establish uniform convergence over suitable increasing sequences of compact sets (cf. Theorem 3.3).

Apart from that, regression and density kernel estimators behave similarly.

For example, under mild conditions we have

$$n^{4/(d+4)} E(r_n(x) - r(x))^2 \longrightarrow c$$

where c is explicit (Theorem 3.1). The proof of this result is rather intricate since it is necessary to use one of the exponential type inequalities established in Chapter 1, in order to control the large deviations of $r_n - r$.

Concerning uniform convergence, a result of the following type may be obtained (Theorem 3.2) :

$$\sup_{x \in S} |r_n(x) - r(x)| = o\left(\frac{(\mathrm{Log} n)^a}{n^{2/(d+4)}}\right) \quad \text{a.s.}$$

where S is a compact set and a is a positive number.

Now, in continuous time, the following result is valid (Corollary 5.1)

$$\limsup_{T \to \infty} \sup_{Z \in \mathcal{Z}} T^{4/(4+d)} E_Z(r_T(x) - r_{(Z)}(x))^2 < \infty$$

where $\mathcal{Z}$ is a suitable family of processes[2]

Similarly, as in the density case, if the sample paths are irregular enough the kernel estimator exhibits a parametric asymptotic behaviour, namely

$$T \cdot E(r_T(x) - r(x))^2 \longrightarrow c$$

where c is explicit (Theorem 5.3).

Finally it may be proved that r_n and r_T have a limit in distribution which is Gaussian (cf. Theorem 3.4 and 5.5 and Corollary 5.2).

---

[2] E_Z denotes the expectation with respect to P_Z and $r_{(Z)}$ the regression associated with Z.

S.6.2   Prediction

The asymptotic properties of the predictors $\widehat{\xi}_{n+H}$ and $\widehat{\xi}_{T+H}$ introduced in S.3 heavily depend on these of the regression estimators which generate them. Details are given in Chapters 3 and 5.

Here we only indicate two noticeable results which are valid under a φ_{rev}-mixing condition (a condition stronger than α-mixing).

Firstly $\widehat{\xi}_{n+H}$ is asymptotically normal and consequently one may construct a confidence interval for ξ_{n+H} (Theorem 3.7).

Secondly, modifying slightly $\widehat{\xi}_{T+H}$ one obtains a new predictor, say ξ^*_{T+H} such that for each compact interval Δ

$$E\left(\xi^*_{T+H} - r(\xi_T)\right)^2 \mathbf{1}_{\xi_T \in \Delta} = O\left(\frac{1}{T}\right)$$

thus the nonparametric predictor ξ^*_{T+H} reaches a parametric rate. This could be a first explanation for the efficiency of nonparametric prediction methods. Other explanations are given in the next section.

S.7   Implementation of nonparametric method

S.7.1   Stationarization

The first step of implementation consists in transformations of the data in order to obtain stationarity. This can be performed by removing trend and seasonality after a preliminary estimation (cf. 3.5.2).

However, the above technique suffers the drawback of perturbating the data. Thus it should be better to use simple transformations as differencing (cf. 3.5.2) or affine transformations (cf. [PO]).

In fact it is even possible to consider directly the original data and use them for prediction! For example if $(\xi_n,\ n \in \mathbb{Z})$ is a real square integrable Markov process, the predictor $\widehat{\xi}_{n+H}$ introduced in S.3 may be written as

$$\widehat{\xi}_{n+H} = \sum_{i=1}^{n-H} p_{in} \xi_{i+H}$$

$$\text{where } p_{in} = \frac{K\left(\dfrac{\xi_n - \xi_i}{h_n}\right)}{\displaystyle\sum_{i=1}^{n-H} K\left(\dfrac{\xi_n - \xi_i}{h_n}\right)} \quad ; \; i = 1, \ldots, n$$

thus $\widehat{\xi}_{n+H}$ is a weighted mean and the weight p_{in} appears as a measure of **similarity** (cf. [PO]) between (ξ_i, ξ_{i+H}) and (ξ_n, ξ_{n+H}). In other words the nonparametric predictor is constructed from the "story" of the process (ξ_t). Consequently trend and seasonality may be used to "tell this story".

Asymptotic mathematical results related to that observation appear in subsection 3.4.2 (see Theorem 3.8).

S.7.2 Construction

The construction of a kernel estimator (or predictor) requires a choice of K and h_n. Some theoretical results show that the choice of reasonable K does not much influence the asymptotic behaviour of f_n or r_n.

On the contrary the choice of h_n turns to be crucial for the estimator's accuracy. Some indications about this choice are given in subsection 3.5.3.

Note that, if the observed random variables are one-dimensional, the normal kernel $K(x) = \dfrac{1}{\sqrt{2\pi}} e^{-\frac{x^2}{2}}$ and $h_n = \widehat{\sigma}_n \, n^{-1/5}$ (where $\widehat{\sigma}_n$ denotes the empirical standard deviation) are commonly used in practice (cf. the appendix).

S.7.3 Sampling

The problem of sampling a continuous time process is considered in Sections 4.4 and 5.6.

The most important concept is "admissible sampling" : given a process $(X_t, \; t \in \mathbb{R})$ with irregular paths, we have seen that superoptimal rates are achieved by nonparametric estimators. For such a process we will say that a sampling is admissible if it corresponds to the minimal number of data preserving the superoptimal rate (in mean square or uniformly).

Theorem 4.12 and 4.13 state that if $X_{\delta_n}, X_{2\delta_n}, \ldots, X_{n\delta_n}$ are observed (with $\delta_n \to 0$ and $T_n = n\delta_n \to \infty$) then $\delta_n = T_n^{-d/2r}$ is admissible provided $h_n = T_n^{-1/2r}$.

S.7.4   Advantages of nonparametric methods

One may summarize the advantages of nonparametric methods as follows :

1) They are robust,

2) Deseasonalization of data is not necessary,

3) In some situations parametric rates are achieved.

Now we do not pretend that the nonparametric kernel method is a "panacea". In discrete time, general "adaptive" methods may be considered (cf. [BI]-[MA] for the i.i.d. case). In continuous time, a new method is considered in [BO] where continuous time processes are interpreted as infinite dimensional autoregressive processes. Semiparametric techniques are also of interest (see for example [RB-ST]).

Concerning the near future of nonparametric we finally enumerate some important topics : study of orthogonal series estimators and predictors (in particular wavelets), image reconstruction, errors in data, presence of exogeneous variables, sampling, estimation and prediction in large dimension. . ..

Chapter 1

Inequalities for mixing processes

In this chapter we present some inequalities for covariances, joint densities and partial sums of stochastic discrete time processes when dependence is measured by strong mixing coefficients. The main tool is coupling with independent random variables. Some limit theorems for mixing processes are given as applications.

1.1 Mixing

In the present paragraph we point out some results about mixing. For the proofs and details we refer to the bibliography.

Let $(\Omega, \mathcal{A}, P)$ be a probability space and let $\mathcal{B}$ and $\mathcal{C}$ be two sub σ-field of $\mathcal{A}$. In order to estimate the correlation between $\mathcal{B}$ and $\mathcal{C}$ various coefficients are used :

- $\alpha = \alpha(\mathcal{B}, \mathcal{C}) = \sup_{\substack{B \in \mathcal{B} \\ C \in \mathcal{C}}} |P(B \cap C) - P(B)P(C)|$,

- $\beta = \beta(\mathcal{B}, \mathcal{C}) = E \sup_{C \in \mathcal{C}} |P(C) - P(C|\mathcal{B})|$,

- $\varphi = \varphi(\mathcal{B}, \mathcal{C}) = \sup_{\substack{B \in \mathcal{B}, \ P(B) > 0 \\ C \in \mathcal{C}}} |P(C) - P(C \mid B)|$,

- $\rho = \rho(\mathcal{B}, \mathcal{C}) = \sup_{\substack{X \in L^2(\mathcal{B}) \\ Y \in L^2(\mathcal{C})}} |corr(X, Y)|$.

These coefficients satisfy the following inequalities :

(1.1) $2\alpha \le \beta \le \varphi$

(1.2) $4\alpha \le \rho \le 2\varphi^{1/2}$.

Now a process $(X_t, t \in \mathbb{Z})$ is said to be **α-mixing** (or **strongly mixing**) if

$$\alpha_k = \sup_{t \in \mathbb{Z}} \alpha(\sigma(X_s, s \le t), \sigma(X_s, s \ge t + k)) \xrightarrow[k \to +\infty]{} 0$$

where the "sup" may be omitted if (X_t) is stationary. Similarly one defines **β-mixing** (or **absolute regularity**), **φ-mixing** and **ρ-mixing**.

By (1.1) and (1.2) we have the following scheme :

$$
\begin{array}{ccc}
\varphi\text{-mixing} & \Longrightarrow & \beta\text{-mixing} \\
\Downarrow & & \Downarrow \\
\rho\text{-mixing} & \Longrightarrow & \alpha\text{-mixing}
\end{array} \quad .
$$

It can be shown that the converse implications do not take place.

As an **example**, consider the linear process

(1.3) $X_t = \sum_{j=0}^{+\infty} a_j \varepsilon_{t-j}$, $t \in \mathbb{Z}$

where $a_j = O(e^{-rj})$, $r > 0$ and where the ε_t's are independent zero-mean real random variables with a common density and finite second moment. Then the series above converges in quadratic mean, and (X_t) is ρ-mixing and therefore α-mixing with coefficients which decrease to zero at an exponential rate.

The existence of a density for ε_t is crucial as the well known following **example** shows : consider the process

$$X_t = \sum_{j=0}^{+\infty} 2^{-j-1} \varepsilon_{t-j}$$, $t \in \mathbb{Z}$

where the ε_t's are independent with common distribution $\mathcal{B}\left(1, \frac{1}{2}\right)$.

Noting that X_t has the uniform density over $(0, 1)$ and that

$$2X_{t+1} = X_t + \varepsilon_{t+1}$$

one deduces that X_t is the fractional part of $2X_{t+1}$, hence $\sigma(X_t) \subset \sigma(X_{t+1})$. By iteration we get

$$\sigma(X_t) \subset \sigma(X_s, s \geq t + k)$$

thus

$$\frac{1}{4} \geq \alpha_k \geq \alpha(\sigma(X_t), \sigma(X_t)) = \frac{1}{4}$$

which proves that (X_t) is not α-mixing. ■

In the **Gaussian case** there are special implications between the various kinds of mixing : if (X_t) is a Gaussian stationary φ-mixing process, then it is **m-dependent** i.e., for some m, $\sigma(X_s, s \leq t)$ and $\sigma(X_s, s \geq t + k)$ are independent for $k > m$. On the other hand we have $\rho_k \leq 2\pi\alpha_k$ for any Gaussian process so that α-mixing and ρ-mixing are equivalent in this particular case. However a Gaussian process may be α-mixing without being β-mixing.

The above results show that φ-mixing and β-mixing are often too restrictive as far as applications are concerned. Further on we will principally use α and ρ-mixing conditions and sometimes the **2-α-mixing condition** :

(1.4) $$\alpha_k^{(2)} = \sup_{t \in \mathbb{Z}} \alpha\left(\sigma(X_t), \sigma(X_{t+k})\right) \xrightarrow[k \to +\infty]{} 0 \ .$$

This condition is weaker than strongly mixing except for a Markov process for which those two mixing conditions coincide.

1.2   Coupling

The use of coupling is fruitful for the study of weakly dependent random variables. The principle is to replace these by independent ones having respectively the same distribution. The difference of behaviour between the two kinds of variable is connected with the mixing coefficients of the dependent random variables. We now state two important coupling results. For the proofs, which are rather intricate, we refer to [BE] and [BR1].

LEMMA 1.1 *(Berbee's lemma)*
  Let (X, Y) be a $\mathbb{R}^d \times \mathbb{R}^{d'}$-valued random vector. Then there exists a $\mathbb{R}^{d'}$-valued random vector Y^* such that

  (1) $P_{Y^*} = P_Y$ and Y^* is independent of X,

  (2) $P(Y^* \neq Y) = \beta(\sigma(X), \sigma(Y))$.

It can be proved that "=" cannot be replaced by "<", thus the result is optimal.

LEMMA 1.2 *(Bradley's lemma)*

   Let (X, Y) be a $\mathbb{R}^d \times \mathbb{R}$-valued random vector such that $Y \in L^p(P)$ for some $p \in [1, +\infty]$. Let c be a real number such that $\|Y+c\|_p > 0$, and $\xi \in]0, \|Y+c\|_p]$. Then, there exists a random variable Y^* such that

   (1) $P_{Y^*} = P_Y$ and Y^* is independent of X,

   (2) $P(|Y^* - Y| > \xi) \leq 11 \left(\xi^{-1} \|Y + c\|_p \right)^{p/(2p+1)} [\alpha(\sigma(X), \sigma(Y))]^{2p/(2p+1)}$.

In the original statement of this lemma, 11 is replaced by 18 and $c = 0$ but the proof is not different. We will see the usefulness of Lemma 1.2 in Section 1.4.

1.3   Inequalities for covariances and joint densities

Essential to the study of estimator's quadratic error are covariance inequalities. The following Rio's inequality is optimal up to a constant factor.

THEOREM 1.1 *(Rio's inequality)*

   Let X and Y be two integrable real-valued random variables and let $Q_X(u) = inf\{t : P(|X| > t) \leq u\}$ be the quantile function of $|X|$. Then if $Q_X Q_Y$ is integrable over $(0, 1)$ we have

$$(1.5) \qquad\qquad |\mathrm{Cov}(X, Y)| \leq 2 \int_0^{2\alpha} Q_X(u) Q_Y(u) du$$

where $\alpha = \alpha(\sigma(X), \sigma(Y))$.

Proof

   Putting $X^+ = sup(0, X)$ and $X^- = sup(0, -X)$ we get

$$(1.6) \qquad \mathrm{Cov}(X, Y) \;=\; \mathrm{Cov}(X^+, Y^+) + \mathrm{Cov}(X^-, Y^-)$$
$$-\mathrm{Cov}(X^-, Y^+) - \mathrm{Cov}(X^+, Y^-).$$

An integration by parts shows that

$$\mathrm{Cov}(X^+, Y^+) = \int_{\mathbb{R}^2_+} [P(X > u, Y > v) - P(X > u)P(Y > v)] du dv,$$

which implies

$$(1.7) \qquad \mathrm{Cov}(X^+, Y^+) \leq \int_{\mathbb{R}^2_+} \inf(\alpha, P(X > u), P(Y > v)) du dv.$$

Now apply (1.6), (1.7) and the elementary inequality
$$(\alpha \wedge a \wedge c) + (\alpha \wedge a \wedge d) + (\alpha \wedge b \wedge c) + (\alpha \wedge b \wedge d) \leq 2[(2\alpha) \wedge (a+b) \wedge (c+d)]$$

to $a = P(X > u)$, $b = P(-X > u)$, $c = P(Y > v)$, $d = P(-Y > v)$ to
obtain

$$|\text{Cov}(X,Y)| \leq 2 \int_{\mathbb{R}^2_+} \inf(2\alpha, P(|X| > u) , P(|Y| > v) du dv =: \mathcal{I}.$$

It remains to prove that

(1.8)
$$\mathcal{I} = 2 \int_0^{2\alpha} Q_X(u) Q_Y(u) du.$$

For that purpose consider a r.v. U with uniform distribution over $[0,1]$ and a
bivariate r.v. (Z,T) defined by

$$(Z,T) = (0,0)1_{U \geq 2\alpha} + (Q_X(U), Q_Y(U))1_{U < 2\alpha}.$$

Thus

$$E(ZT) = \int_0^{2\alpha} Q_X(u) Q_Y(u) du$$

and

$$(Z > u, T > v) = (U < 2\alpha, U < P(|X| > u), U < P(|Y| > v)) ,$$

hence

$$\begin{aligned} E(ZT) &= \int_{\mathbb{R}^2_+} P(Z > u, T > v) du dv \\ &= \int_{\mathbb{R}^2_+} \inf(2\alpha, P(|X| > u), P(|Y| > v)) du dv \end{aligned}$$

which entails (1.8) and the proof is thus complete. ∎

Conversely it can be proved that if μ is a symmetric probability distribu-
tion over $\mathbb{R}$ and if $\alpha \in]0, \frac{1}{4}]$, there exists two r.v.'s X and Y with common
distribution μ such that $\alpha(\sigma(X), \sigma(Y)) \leq \alpha$ and

(1.9)
$$\text{Cov}(X,Y) \geq \frac{1}{2} \int_0^{2\alpha} [Q_X(u)]^2 du .$$

Proof may be found in [RI].

We now present two inequalities which are less general but more tractable.

COROLLARY 1.1 *Let X and Y be two real valued random variables such
that $X \in L^q(P)$, $Y \in L^r(P)$ where $q > 1$, $r > 1$ and $\frac{1}{q} + \frac{1}{r} = 1 - \frac{1}{p}$, then*

(1.10)
$$|\text{Cov}(X,Y)| \leq 2p(2\alpha)^{1/p} \|X\|_q \|Y\|_r$$

(Davydov's inequality).

In particular if $X \in L^\infty(P)$, $Y \in L^\infty(P)$ then

(1.11) $|\text{Cov}(X, Y)| \leq 4\|X\|_\infty \|Y\|_\infty \, \alpha$

(Billingsley's inequality).

Proof

Suppose first that q and r are finite. Then Markov's inequality yields

$$P\left(|X| > \frac{\|X\|_q}{u^{1/q}}\right) \leq u \quad , \quad 0 < u \leq 1$$

which implies

$$Q_X(u) \leq \frac{\|X\|_q}{u^{1/q}} \quad , \quad 0 < u \leq 1.$$

Now, using (1.5) we obtain

$$|\text{Cov}(X, Y)| \quad \leq 2 \int_0^{2\alpha} \frac{\|X\|_q}{u^{1/q}} \frac{\|Y\|_r}{u^{1/r}} du$$

$$\leq 2\|X\|_q \|Y\|_r \int_0^{2\alpha} u^{\frac{1}{p}-1} du$$

hence (1.10).

If $q = r = +\infty$ we clearly have

$$Q_X(u) \leq Q_X(0) = \|X\|_\infty$$

thus

$$2\int_0^{2\alpha} Q_X(u) Q_Y(u) du \leq 4\alpha \|X\|_\infty \|Y\|_\infty . \blacksquare$$

Note that (1.10) is valid if $q = +\infty$ and $r > 1$. If $q = 1$ or $r = 1$ the resulting inequality becomes trivial.

We now consider the local measure of dependence defined by

(1.12) $g_{(X,Y)}(x, y) = f_{(X,Y)}(x, y) - f_X(x) f_Y(y) \; ; \; x, y \in \mathbb{R}^d,$

where (X, Y) is a $\mathbb{R}^d \times \mathbb{R}^d$-valued random vector and where f_Z denotes the density of the random vector Z with respect to Lebesgue measure.

The following statement connects $g = g_{(X,Y)}$ with $\alpha = \alpha(\sigma(X), \sigma(Y))$.

LEMMA 1.3 *If (X, Y) has an absolutely continuous distribution with respect to Lebesgue measure on $\mathbb{R}^{2d}$ then*

(1.13)
$$\alpha \leq \frac{1}{2}\|g\|_1.$$

If in addition g satisfies the Lipschitz's condition

(1.14)
$$|g(x', y') - g(x, y)| \leq \ell(\|x' - x\|^2 + \|y' - y\|^2)^{1/2},$$

$x, x', y, y' \in \mathbb{R}^d$, *for some constant ℓ, then there exists a constant $\gamma(d, \ell)$ such that*

(1.15)
$$\|g\|_\infty \leq \gamma(d, \ell)\alpha^{1/(2d+1)}.$$

Furthermore one may choose $\gamma(d, \ell) = V_d^{-2} + \ell\sqrt{2}$ where V_d denotes the volume of the unit ball in $\mathbb{R}^d$.

Proof

By the definition of α it is clear that

$$\alpha \leq \sup_{B \in \mathcal{B}_{\mathbb{R}}^{2d}} |P_{(X,Y)}(B) - (P_X \otimes P_Y)(B)| =: S.$$

Now using Scheffe's theorem (cf. [BI1] p. 224) we obtain

$$S = \frac{1}{2}\|g\|_1$$

hence (1.13).

On the other hand, set

$$B(x, \varepsilon) = \{x' : \|x' - x\| \leq \varepsilon\} \ , \ \varepsilon > 0, \ x \in \mathbb{R}^d.$$

Then, for any $(x, y) \in \mathbb{R}^{2d}$, we have

$$\alpha \ \geq |P(X \in B(x, \varepsilon), Y \in B(y, \varepsilon)) - P(X \in B(x, \varepsilon))P(Y \in B(y, \varepsilon))|$$

$$\geq \left|\int_{B(x,\varepsilon) \times B(y,\varepsilon)} g(u, v)dudv\right| =: \mathcal{I}.$$

Now by the mean value property we get

$$\mathcal{I} = V_d^2 \varepsilon^{2d}|g(x', y')|$$

for some (x', y') in $B(x, \varepsilon) \times B(y, \varepsilon)$.

On the other hand (1.14) yields

$$|g(x, y)| \leq |g(x', y')| + \ell\varepsilon\sqrt{2}$$

hence

$$|g(x, y)| \leq \frac{\alpha}{V_d^2 \varepsilon^{2d}} + \ell\varepsilon\sqrt{2}$$

choosing $\varepsilon = \alpha^{1/(2d+1)}$ we obtain (1.15). ∎

1.4   Exponential type inequalities

We now turn to the study of large deviations for partial sums of strongly mixing processes.

   Let us begin with exponential type inequalities for independent random variables.

THEOREM 1.2 *Let $X_1, \ldots, X_n$ be independent zero-mean real-valued random variables and let $S_n = \sum\limits_{i=1}^{n} X_i$. The following inequalities hold*

(1) If $a_i \leq X_i \leq b_i$; $i = 1, \ldots, n$ where $a_1, b_1, \ldots, a_n, b_n$ are constant then

$$(1.16) \qquad P(|S_n| \geq t) \leq 2 \exp\left(-\frac{2t^2}{\sum\limits_{i=1}^{n}(b_i - a_i)^2} \right) \, , \, t > 0$$

(Hoeffding's inequality).

(2) If there exists $c > 0$ such that

$$(1.17) \qquad\qquad E|X_i|^p \leq c^{p-2} p! EX_i^2 < +\infty$$

$i = 1, \ldots, n$; $p = 3, 4, \ldots$
(Cramer's conditions) then

$$(1.18) \qquad P(|S_n| \geq t) \leq 2 \exp\left(-\frac{t^2}{4 \sum\limits_{i=1}^{n} EX_i^2 + 2ct} \right) \, , t > 0$$

(Bernstein's inequality).

Proof

(1) First, let X be a real-valued zero-mean random variable such that $a \leq X \leq b$. We claim that

$$(1.19) \qquad E(\exp \lambda X) \leq \exp\left(\frac{\lambda^2 (b-a)^2}{8} \right) \, , \lambda > 0.$$

In order to prove (1.19) we consider the convexity inequality

$$e^{\lambda x} \leq \frac{b-x}{b-a} e^{\lambda a} + \frac{x-a}{b-a} e^{\lambda b} \, , a \leq x \leq b.$$

Replacing x by X and taking the expectation, it follows that

$$E(e^{\lambda X}) \le \frac{b}{b-a}e^{\lambda a} - \frac{a}{b-a}e^{\lambda b} =: \varphi.$$

Thus

$$\varphi = [1 - p + pe^{\lambda(b-a)}]e^{-p\lambda(b-a)}$$
$$=: \exp(\psi(u))$$

where $p = \dfrac{a}{a-b}$, $u = \lambda(b-a)$, $\psi(u) = -pu + \text{Log}(1 - p + pe^u)$.
Now it is easy to check that $\psi(0) = \psi'(0) = 0$ and

$$\psi''(u) = \frac{p(1-p)e^{-u}}{[p + (1-p)e^{-u}]^2} \le \frac{1}{4},$$

consequently the Taylor's formula leads to

$$\psi(u) \le \frac{u^2}{8} = \frac{\lambda^2(b-a)^2}{8}$$

hence (1.19). ∎

We are now in a position to establish (1.16).
The main tool is the famous "Bernstein's trick" : since

(1.20) $$1_{S_n \ge t} \le e^{\lambda(S_n - t)} \quad, \lambda > 0$$

we have

$$P(S_n \ge t) \le e^{-\lambda t} E(e^{\lambda S_n})$$
$$\le e^{-\lambda t} \prod_{i=1}^{n} E(e^{\lambda X_i}).$$

Now applying (1.19) to $X_1, \dots, X_n$ we obtain

$$P(S_n \ge t) \le e^{-\lambda t} \exp\left(+\frac{\lambda^2 \sum_{i=1}^{n}(b_i - a_i)^2}{8}\right).$$

Choosing $\lambda = \dfrac{4t}{\sum\limits_{i=1}^{n}(b_i - a_i)^2}$ it follows that

$$P(S_n \ge t) \le \exp\left(-\frac{2t^2}{\sum\limits_{i=1}^{n}(b_i - a_i)^2}\right) =: A.$$

Similarly an application of (1.19) to the random variables $-X_i$ shows that

$$P(S_n \leq -t) = P(-S_n \geq t) \leq A$$

and the proof is complete since

(1.21) $$P(|S_n| \geq t) = P(S_n \geq t) + P(S_n \leq -t). \quad \blacksquare$$

(2) For $0 < \lambda < \dfrac{1}{c}$ according to Cramer's conditions (1.17) we have

(1.22) $$\sum_{p=2}^{+\infty} \frac{1}{p!} E(|\lambda X_i|^p) \leq \lambda^2 EX_i^2 \sum_{p=2}^{+\infty} (\lambda c)^{p-2} = \frac{\lambda^2 EX_i^2}{1 - \lambda c}.$$

Using (1.22) and the dominated convergence theorem we can deduce that

$$
\begin{aligned}
E(e^{\lambda X_i}) &\leq 1 + \sum_{p=2}^{+\infty} \frac{\lambda^p E(|X_i|^p)}{p!} \\
&\leq 1 + \frac{\lambda^2 EX_i^2}{1 - \lambda c} \\
&\leq \exp\left(\frac{\lambda^2 EX_i^2}{1 - \lambda c}\right).
\end{aligned}
$$

Using again the Bernstein's trick we obtain

$$
\begin{aligned}
P(S_n \geq t) &\leq e^{-\lambda t} \prod_{i=1}^{n} E(e^{\lambda X_i}) \\
&\leq e^{-\lambda t} \exp\left(\frac{\lambda^2 \sum_{i=1}^{n} EX_i^2}{1 - \lambda c}\right).
\end{aligned}
$$

Now the choice $\lambda = \dfrac{t}{2 \sum_{i=1}^{n} EX_i^2 + ct}$ leads to

$$P(S_n \geq t) \leq \exp\left(-\frac{t^2}{4 \sum_{i=1}^{n} EX_i^2 + 2ct}\right).$$

and it suffices to use (1.21) to get the desired result. ■

It should be noticed that these inequalities are optimal up to a constant in the exponent as the following Kolmogorov's converse exponential inequality shows : if conditions in Theorem 1.2 (1) hold with $b_i = a_i = b$, $i = 1, \ldots, n$, then, for any $\gamma > 0$ there exist $k(\gamma) > 0$ and $\varepsilon(\gamma) > 0$ such that if

$$t \geq k(\gamma) \left(\sum_{i=1}^n EX_i^2 \right)^{1/2} \quad \text{and} \quad tb \leq \varepsilon(\gamma) \left(\sum_{i=1}^n EX_i^2 \right)$$

it can be inferred that

$$(1.23) \qquad P(S_n \geq t) \geq \exp \left(-\frac{1+\gamma}{2 \sum_{i=1}^n EX_i^2} t^2 \right).$$

We refer to [ST] for a proof of this inequality.

We now turn to the study of the dependent case.
For **any** real discrete time process $(X_t, t \in \mathbb{Z})$ we define the strongly mixing coefficients as

$$(1.24) \qquad \alpha(k) = \sup_{t \in \mathbb{Z}} \alpha \left(\sigma(X_s, s \leq t), \sigma(X_s, s \geq t + k) \right) \; ; k = 1, 2, \ldots.$$

Note that this scheme applies to a finite number of random variables since it is always possible to complete a sequence by adding an infinite number of degenerate random variables.

The following theorem provides inequalities for bounded stochastic processes.

THEOREM 1.3 *Let $(X_t, t \in \mathbb{Z})$ be a zero-mean real-valued process such that $\sup_{1 \leq t \leq n} \|X_t\|_\infty \leq b$. Then*

(1) For each integer $q \in \left[1, \dfrac{n}{2} \right]$ and each $\varepsilon > 0$

$$(1.25) \qquad P(|S_n| > n\varepsilon) \leq 4 \exp \left(-\frac{\varepsilon^2}{8b^2} q \right)$$

$$+ 22 \left(1 + \frac{4b}{\varepsilon} \right)^{1/2} q \alpha \left(\left[\frac{n}{2q} \right] \right).$$

(2) *For each integer $q \in \left[1, \dfrac{n}{2}\right]$ and each $\varepsilon > 0$*

(1.26) $$P(|S_n| > n\varepsilon) \leq 4\exp\left(-\frac{\varepsilon^2}{8v^2(q)}q\right)$$

$$+22\left(1 + \frac{4b}{\varepsilon}\right)^{1/2}q\alpha\left(\left[\frac{n}{2q}\right]\right),$$

where

$$v^2(q) = \frac{2}{p^2}\sigma^2(q) + \frac{b\varepsilon}{2}$$

with $p = \dfrac{n}{2q}$ *and* $\sigma^2(q) = \max\limits_{0 \leq j \leq 2q-1} E\left(([jp] + 1 - jp)X_{[jp]+1} + X_{[jp]+2} + \right.$
$$\left. \ldots + X_{[(j+1)p]} + ((j+1)p - [(j+1)p])\, X_{[(j+1)p+1]}\right)^2.$$

Proof

(1) Consider the auxiliary continuous time process $Y_t = X_{[t+1]}$, $t \in \mathbb{R}$. We clearly have $S_n = \int_0^n Y_u du$.

Let us now define "blocks" as follows

$$V_1 = \int_0^p Y_u du \qquad , \qquad V_1' = \int_p^{2p} Y_u du$$

$$\vdots$$

$$V_q = \int_{2(q-1)p}^{(2q-1)p} Y_u du \quad , \quad V_q' = \int_{(2q-1)p}^{2qp} Y_u du \ ,$$

where $p = \dfrac{n}{2q}$.

Using recursively Bradley's lemma 1.2 we may define independent r.v.'s $W_1, \ldots, W_q$ such that $P_{W_j} = P_{V_j}$, $j = 1, \ldots, q$ and

(1.27) $$P(|W_j - V_j| > \xi) \leq 11\left(\frac{\|V_j + c\|_\infty}{\xi}\right)^{1/2}\alpha([p])$$

Details about the construction of $W_1, \ldots, W_q$ may be found in [TR].

Here $c = \delta b p$ and $\xi = \min\left(\dfrac{n\varepsilon}{4q}, (\delta - 1)bp\right)$ for some $\delta > 1$ which will be specified below.

Note that, for each j,

$$\|V_j + c\|_\infty \geq c - \|V_j\|_\infty \geq (\delta - 1)bp > 0$$

so that $0 < \xi \le \|V_j + c\|_\infty$ as required in Lemma 1.2 .

Now, according to the choice of c and ξ, (1.27) may be written

$$P(|W_j - V_j| > \xi) \quad \le 11 \left(\frac{(\delta + 1)bp}{\min\left((n\varepsilon/(4q)), (\delta - 1)bp\right)} \right)^{1/2} \alpha([p])$$

$$\le 11 \left(\max \left(\frac{\delta + 1}{\delta - 1}, \frac{4qbp(\delta + 1)}{n\varepsilon} \right) \right)^{1/2} \alpha([p]).$$

If $\delta = 1 + \dfrac{\varepsilon}{2b}$ then

$$P(|W_j - V_j| > \xi) \le 11 \left(2 + \frac{\varepsilon}{2b}\right)^{1/2} \left(\frac{2b}{\varepsilon}\right)^{1/2} \alpha([p])$$

thus

(1.28) $$P(|W_j - V_j| > \xi) \le 11 \left(1 + \frac{4b}{\varepsilon}\right)^{1/2} \alpha([p]).$$

On the other hand we may apply Hoeffding's inequality (1.16) to the W_j's. We then obtain

(1.29) $$P\left(\left|\sum_1^q W_j\right| > \frac{n\varepsilon}{4}\right) \le 2\exp\left(-\frac{n\varepsilon^2}{16pb^2}\right).$$

We are now in a position to conclude.

Clearly

(1.30) $$P(|S_n| > n\varepsilon) \le$$

$$P\left(\left|\sum_1^q V_j\right| > \frac{n\varepsilon}{2}\right) + P\left(\left|\sum_1^q V_j'\right| > \frac{n\varepsilon}{2}\right).$$

and

$$\left\{\left|\sum_1^q V_j\right| > \frac{n\varepsilon}{2}\right\} \subset \left\{\left|\sum_1^q V_j\right| > \frac{n\varepsilon}{2} \; ; \; |V_j - W_j| \le \xi \; ; \; j = 1, \ldots, q\right\}$$

$$\cup \left\{\bigcup_1^q |V_j - W_j| > \xi\right\},$$

hence

$$P\left(\left|\sum_1^q V_j\right| > \frac{n\varepsilon}{2}\right) \leq P\left(\left|\sum_1^q W_j\right| > \frac{n\varepsilon}{2} - q\xi\right) + \sum_1^q P(|V_j - W_j| > \xi)$$

$$\leq P\left(\left|\sum_1^q W_j\right| > \frac{n\varepsilon}{4}\right) + \sum_1^q P(|V_j - W_j| > \xi).$$

Consequently (1.28) and (1.29) give the upper bound

$$P\left(\left|\sum_1^q V_j\right| > \frac{n\varepsilon}{2}\right) \leq 2\exp\left(-\frac{\varepsilon^2}{8b^2}q\right) + 11\left(1 + \frac{4b}{\varepsilon}\right)^{1/2} q\alpha([p]),$$

and the same bound is valid for the V_j's. According to (1.30), inequality (1.25) is thus established. ∎

(2) The proof of (1.26) is similar except that, here, we use Bernstein's inequality (1.18) instead of the Hoeffding's one.
So we have

(1.31) $$P\left(\left|\sum_1^q W_j\right| > \frac{n\varepsilon}{4}\right) \leq$$

$$2\exp\left(-\frac{n^2\varepsilon^2/16}{4\sum_1^q EW_j^2 + 2bpn\varepsilon/4}\right).$$

Now, since $P_{W_j} = P_{V_j}$ we have

$$EW_j^2 = EV_j^2 = E\left(\int_{jp}^{(j+1)p} Y_u du\right)^2$$

and

$$E\left(\int_{jp}^{(j+1)p} Y_u du\right)^2 = E\left(([jp] + 1 - jp)X_{[jp]+1} + X_{[jp]+2} + \cdots \right.$$
$$\left. + X_{[(j+1)p]} + ((j+1)p - [(j+1)p]) X_{[(j+1)p]+1}\right)^2.$$

Taking into account the above overestimate and using (1.31) we obtain after some easy calculations

$$P\left(\left|\sum_1^q W_j\right| > \frac{n\varepsilon}{4}\right) \leq 2\exp\left(-\frac{\varepsilon^2 q}{8v^2(q)}\right)$$

which entails (1.26). ∎

Note that by using (1.11) it is easy to see that

$$(1.32) \qquad v^2(q) \leq \frac{2}{p^2} \left(\max_{1 \leq t \leq n} EX_t^2 + 8b^2 \sum_{k=1}^{[p]+1} \alpha(k) \right) + \frac{b\varepsilon}{2}.$$

We would like to mention that although (1.26) is sharper than (1.25) when ε and $\alpha(.)$ are small enough, however (1.25) is more tractable in some practical situations.

The next theorem is devoted to the general case where the X_t's are not necessarily bounded but satisfy Cramer's conditions.

THEOREM 1.4 *Let* $(X_t, t \in \mathbb{Z})$ *be a zero-mean real-valued process. Suppose that there exists $c > 0$ such that*

$$(1.33) \qquad E|X_t|^k \leq c^{k-2} k! EX_t^2 < +\infty \; ; \; t = 1, \ldots, n \; ; \; k = 3, 4, \ldots$$

then for each $n \geq 2$, each integer $q \in \left[1, \dfrac{n}{2}\right]$, each $\varepsilon > 0$ and each $k \geq 3$

$$(1.34) \qquad P(|S_n| > n\varepsilon) \leq$$

$$a_1 \exp\left(-\frac{q\varepsilon^2}{25m_2^2 + 5c\varepsilon}\right) + a_2(k)\alpha\left(\left[\frac{n}{q+1}\right]\right)^{\frac{2k}{2k+1}}$$

where

$$a_1 = 2\frac{n}{q} + 2\left(1 + \frac{\varepsilon^2}{25m_2^2 + 5c\varepsilon}\right), \; with \; m_2^2 = \max_{1 \leq t \leq n} EX_t^2,$$

and

$$a_2(k) = 11n\left(1 + \frac{5m_p}{\varepsilon}^{\frac{k}{2k+1}}\right), \; with \; m_p = \max_{1 \leq t \leq n} \|X_t\|_p.$$

Proof

Let q and r be integers such that

$$1 \leq qr \leq n < (q+1)r.$$

Consider the partial sums

$$
\begin{array}{llllllll}
Z_1 & = & X_1 & + & X_{r+1} & + & \cdots & + & X_{(q-1)r+1} \\
Z_2 & = & X_2 & + & X_{r+2} & + & \cdots & + & X_{(q-1)r+2} \\
\vdots & & \vdots & & \vdots & & & & \vdots \\
Z_r & = & X_r & + & X_{2r} & + & \cdots & + & X_{qr} \\
\Delta & = & X_{qr+1} & + & \cdots & & + & \cdots & + & X_n & \text{if } qr < n \\
& = & 0 & & & & & & & & \text{otherwise.}
\end{array}
$$

We clearly have

$$(1.35) \qquad P(|S_n)] > n\varepsilon) \leq \sum_{j=1}^{r} P\left(|Z_j| > \frac{4n\varepsilon}{5r}\right) + P\left(|\Delta| > \frac{n\varepsilon}{5}\right).$$

Now, in order to get an upper bound for $P\left(|Z_1| > \dfrac{4n\varepsilon}{5r}\right)$ we apply recursively Bradley's lemma 1.2 : let k be an integer ≥ 2, δ a real > 1 and ξ such that

$$0 < \xi \leq (\delta - 1)m_k \leq \|X_{(j-1)r+1} + \delta m_k\|_k \leq (\delta + 1)m_k \; ; \; j = 1, \ldots, q \, .$$

We may and do suppose that m_k is strictly positive, otherwise the inequality should be trivial.

Then, there exist independent r.v.'s Y_j , $j = 1, \ldots, q$ such that $P_{Y_j} = P_{X_{(j-1)r+1}}$ and

$$(1.36) \qquad P(|Y_j - X_{(j-1)r+1}| > \xi) \leq$$

$$11 \left(\frac{\|X_{(j-1)r+1} + \delta m_k\|_k}{\xi} \right)^{\frac{k}{2k+1}} (\alpha(r))^{\frac{2k}{2k+1}} .$$

Choosing

$$\delta = 1 + \frac{2\varepsilon}{5m_k} \quad \text{and} \quad \xi = \frac{2\varepsilon}{5} \text{ yields}$$

$$(1.37) \quad P\left(|Y_j - X_{(j-1)r+1}| > \frac{2\varepsilon}{5}\right) \leq 11 \left(1 + \frac{5m_k}{\varepsilon}\right)^{\frac{k}{2k+1}} (\alpha(r))^{\frac{2k}{2k+1}} .$$

Now elementary computations give

$$(1.38) \qquad P\left(|Z_1| > \frac{4n\varepsilon}{5r}\right) \leq P\left(|Y_1 + \ldots + Y_q| > \frac{2q\varepsilon}{5}\right)$$

$$+ \sum_{j=1}^{q} P\left(|Y_j - X_{(j-1)r+1}| > \frac{2\varepsilon}{5}\right).$$

Applying Bernstein's inequality (1.18) to the Y_j's we obtain

$$(1.39) \qquad P\left(|Y_1 + \ldots + Y_q| > \frac{2q\varepsilon}{5}\right) \leq 2\exp\left(-\frac{q\varepsilon^2}{25m_2^2 + 5c\varepsilon}\right).$$

Thus combining (1.37), (1.38) and (1.39) we get an upper bound for $P\left(|Z_1| > \dfrac{4n\varepsilon}{5r}\right)$. Clearly the same bound remains valid for $Z_2, \ldots, Z_r$.

The proof will be complete if we exhibit a suitable overestimate for $P\left(|\Delta| > \frac{n\varepsilon}{5}\right)$.

For that purpose we write

$$
\begin{aligned}
P\left(\Delta > \frac{n\varepsilon}{5}\right) &\leq \exp(-\lambda n\varepsilon/5)E(e^{\lambda\Delta}) \ , \ \lambda > 0 \\
&\leq \exp(-\lambda n\varepsilon/5)\left(1 + \sum_{k=2}^{\infty} \frac{\lambda^k}{k!}E|\Delta|^k\right).
\end{aligned}
$$

Now Minkowski's inequality and (1.33) entail

$$
\begin{aligned}
E|\Delta|^k &\leq \left(\|X_{qr+1}\|_k + \ldots + \|X_n\|_k\right)^k \\
&\leq (n - qr)^k c^{k-2} k! m_2^2 \ , \ k \geq 2.
\end{aligned}
$$

Hence for a suitable λ

$$
P\left(\Delta > \frac{n\varepsilon}{5}\right) \leq \exp(-\lambda n\varepsilon/5)\left(1 + \lambda^2(n - qr)^2 m_2^2 \sum_{k=2}^{\infty}(\lambda c(n - qr))^{k-2}\right),
$$

thus choosing $\lambda = \theta/(n - qr)c$, $0 < \theta < 1$ we get

$$
P\left(\Delta > \frac{n\varepsilon}{5}\right) \leq \left(1 + \frac{\theta^2}{c^2}\frac{m_2^2}{1 - \theta}\right)\exp\left(\frac{\theta\varepsilon}{5c}\frac{n}{r}\right).
$$

Using the same method for $-\Delta$ we obtain

$$
P\left(|\Delta| > \frac{n\varepsilon}{5}\right) \leq 2\left(1 + \frac{\theta^2}{c^2}\frac{m_2^2}{1 - \theta}\right)\exp\left(-\frac{\theta\varepsilon}{5c}q\right).
$$

Choosing $\theta = c\varepsilon/(5m_2^2 + c\varepsilon)$ yields

$$
P\left(|\Delta| > \frac{n\varepsilon}{5}\right) \leq 2\left(1 + \frac{\varepsilon^2}{5(5m_2^2 + c\varepsilon)}\right)\exp\left(-\frac{\varepsilon^2}{5(5m_2^2 + c\varepsilon)}q\right).
$$

Collecting the above bounds we obtain the claimed result according to (1.35).
∎

1.5 Some limit theorems for strongly mixing processes

It is well known that the laws of large numbers hold for stochastic processes provided classical ergodicity conditions (cf. [DO]). However the Statistician needs some convergence rate in order to convince himself of applicability of the theoretical results.

The present section is devoted to the study of convergence rates under strongly mixing conditions. To this aim we use the inequalities in the previous sections.
We first state a result concerning the weak law of large numbers.

THEOREM 1.5 *Let $(X_t, t \in \mathbb{Z})$ be a zero-mean real-valued stationary process such that for some $r > 2$*

$$\sup_{t \in \mathbb{Z}} E|X_t|^r < +\infty$$

and

$$\sum_{k \geq 1} \alpha(k)^{1-\frac{2}{r}} < +\infty$$

then the series $\sum_{k \in \mathbb{Z}} \mathrm{Cov}(X_0, X_k)$ is absolutely convergent, has a nonnegative sum σ^2 and

(1.40)
$$n \mathrm{Var} \frac{S_n}{n} \to \sigma^2.$$

Proof

First we study the series $\sum_{k \in \mathbb{Z}} \mathrm{Cov}(X_0, X_k)$. By using (1.10) with $q = r$ and $\frac{1}{p} = 1 - \frac{2}{r}$ we get

$$|\mathrm{Cov}(X_0, X_k)| \leq 2 \frac{r}{r-2} (2\alpha(k))^{1-2/r} (E|X_0|^r)^{2/r}$$

which proves the absolute convergence of the series since $\sum_{k \geq 1} \alpha(k)^{1-2/r} < +\infty$.

Now clearly

$$n \mathrm{Var} \frac{S_n}{n} = n^{-1} \sum_{0 \leq s,t \leq n-1} \mathrm{Cov}(X_s, X_t),$$

(X_t) being stationary it follows that

$$n \mathrm{Var} \frac{S_n}{n} = \sum_{k=-(n-1)}^{n-1} \left(1 - \frac{|k|}{n}\right) \mathrm{Cov}(X_0, X_k).$$

Thus an application of the Lebesgue dominated convergence theorem entails

$$\lim_{n \to \infty} n \mathrm{Var} \frac{S_n}{n} = \sigma^2 \geq 0$$

and the theorem is thus established. ∎

The following proposition provides pointwise results.

THEOREM 1.6 *Let $(X_t, t \in \mathbb{Z})$ be a zero-mean real-valued process satisfying Cramer's conditions (1.33). We have the following*

(1) *If (X_t) is m-dependent, then*

(1.41)
$$\frac{S_n}{\sqrt{n\mathrm{Log}_2 n\mathrm{Log}n}} \to 0 \quad a.s. \ .$$

(2) *If (X_t) is α-mixing with $\alpha(k) \leq a\rho^k$, $a > 0$, $0 < \rho < 1$ then*

(1.42)
$$\frac{S_n}{\sqrt{n\mathrm{Log}_2 n\mathrm{Log}n}} \to 0 \quad a.s. \ .$$

Proof

(1) Using (1.34) for $n > m$, $\varepsilon = \sqrt{\dfrac{\mathrm{Log}_2 n\mathrm{Log}n}{n}}\eta$, $\eta > 0$ and $q = [n/m + 1]$ we get

$$P\left(\frac{|S_n|}{\sqrt{n\mathrm{Log}_2 n\mathrm{Log}n}} > \eta\right)$$

$$\leq \left(4(m + 1) + 2\left(1 + O\left(\frac{\mathrm{Log}_2 n.\mathrm{Log}n}{n}\right)\right)\right)\exp(-d\mathrm{Log}_2 n\mathrm{Log}n)$$

where d is some positive constant. Therefore

$$\sum_{n>m} P\left(\frac{|S_n|}{\sqrt{n\mathrm{Log}_2 n\mathrm{Log}n}} > \eta\right) < +\infty \ , \ \eta > 0$$

and the Borel Cantelli lemma (cf. [BI 2]) yields (1.41). ∎

(2) Using again (1.34) with $\varepsilon = \dfrac{\mathrm{Log}_2 n\mathrm{Log}n}{\sqrt{n}}\eta$, $\eta > 0$, $k = 2$ and $q = \left[\dfrac{n}{\mathrm{Log}_2 n\mathrm{Log}n} + 1\right]$ leads to

$$P\left(\frac{|S_n|}{\sqrt{n}\mathrm{Log}_2 n\mathrm{Log}n} > \eta\right) =$$
$$O(\mathrm{Log}_2 n\mathrm{Log}n \exp(-d'\mathrm{Log}_2 n\mathrm{Log}n)) + O(n\exp(-d''\mathrm{Log}_2 n\mathrm{Log}n))$$

where d' and d'' are some positive constant. Hence (1.42) using again Borel-Cantelli lemma. ∎

Note that (1.41) and (1.42) are nearly optimal since the law of the iterated logarithm implies that $\dfrac{S_n}{\sqrt{n \text{Log}_2 n}} \not\to 0$ a.s. even for independent summands.

We now give a central limit theorem for strongly mixing processes.

THEOREM 1.7 *Suppose that $(X_t, t \in \mathbb{Z})$ is a zero-mean real-valued strictly stationary process such that for some $\gamma > 2$ and some $\delta > 0$*

$$E|X_t|^\gamma < +\infty$$

and

(1.43)
$$\alpha(k) \le ak^{-\beta}$$

where a is a positive constant and $\beta > \dfrac{\gamma}{\gamma - 2}$,

then, if $\sigma^2 = \displaystyle\sum_{k=-\infty}^{+\infty} \text{Cov}(X_0, X_k) > 0$ we have

(1.44)
$$\frac{S_n}{\sigma\sqrt{n}} \xrightarrow{W} N \sim \mathcal{N}(0, 1).$$

Proof

First σ^2 does exist by Theorem 1.5. Now consider the blocks

$$V_1 = X_1 + \ldots + X_p \qquad\qquad V_1' = X_{p+1} + \ldots + X_{p+q}$$
$$\vdots \qquad\qquad\qquad\qquad\qquad \vdots$$
$$V_r = X_{(r-1)(p+q)+1} + \ldots + X_{rp+(r-1)q} \qquad V_r' = X_{rp+(r-1)q+1} + \ldots + X_{r(p+q)}$$

where

$$r(p + q) \le n < r(p + q + 1)$$

and $r \sim \text{Log} n$, $p \sim \dfrac{n}{\text{Log} n} - n^{1/4}$, $q \sim n^{1/4}$.

Using Lemma 1.2 we construct independent random variables $W_1, \ldots, W_r$ such that $P_{W_j} = P_{V_j}$ and

(1.45)
$$P(|W_j - V_j| > \xi) \le 11 \left(\frac{\|V_j + c\|_\gamma}{\xi} \right)^{\frac{\gamma}{2\gamma+1}} \alpha(q)^{\frac{\gamma}{2\gamma+1}} ;$$

$j = 1, \ldots, r$; where $\xi = \dfrac{\varepsilon\sigma\sqrt{n}}{r}$ $(\varepsilon > 0)$, $c = p\,\zeta\|X_0\|_\gamma$ $(\zeta > 1)$.

Note that for n large enough we have

$$\|V_j + c\|_\gamma \ge c - \|V_j\|_\gamma \ge (\zeta - 1)p\|X_0\|_\gamma \ge \frac{\varepsilon}{\sigma\sqrt{n}r}$$

since $p \sim \dfrac{n}{Logn}$ and $\dfrac{\sqrt{n}}{r} \sim \dfrac{\sqrt{n}}{Logn}$, so that (1.45) is valid.

Consequently setting

$$\Delta_n = \frac{V_1 + \ldots + V_r}{\sigma\sqrt{n}} - \frac{W_1 + \ldots + W_r}{\sigma\sqrt{n}}$$

we obtain

$$P(|\Delta_n| > \varepsilon) \le \sum_{j=1}^{r} P\left(|V_j - W_j| > \frac{\varepsilon\sigma\sqrt{n}}{r}\right)$$

thus

(1.46) $$P(|\Delta_n| > \varepsilon) \le 11r \left(\frac{\|V_1 + c\|_\gamma}{\xi}\right)^{\frac{\gamma}{2\gamma+1}} \alpha(q)^{\frac{2\gamma}{2\gamma+1}} =: m_n.$$

Now let us prove the asymptotic normality of $\dfrac{W_1 + \ldots + W_r}{\sigma\sqrt{n}}$. First using (1.43) and combinatorial arguments it can be checked that for $2 < \gamma' < \gamma$ and γ' enough close to 2

$$E|W_j|^{\gamma'} \le \eta p^{\gamma'/2} \quad ; j = 1, \ldots, r$$

where η is a positive constant. We refer to [YO] for the details.

On the other hand, using stationarity and (1.40) we get

$$EW_j^2 = EV_j^2 = ES_p^2 \sim \sigma^2 p.$$

We are now in a position to show that Liapounov's condition (see [BI2]) holds. Actually

$$\sum_{j=1}^{r} \frac{E|W_j|^{\gamma'}}{\left(Var\sum_{1}^{r} W_j\right)^{\gamma'/2}} = O\left(r\frac{\eta p^{\gamma'/2}}{(\sigma^2 rp)^{\gamma'/2}}\right) = O\left(r^{1-\frac{\gamma'}{2}}\right)$$

and $r^{1-\frac{\gamma'}{2}} \to 0$ since $\gamma' > 2$. Consequently

$$\frac{W_1 + \ldots + W_r}{\sigma\sqrt{n}} = \left(\frac{rp}{n}\right)^{1/2} \frac{W_1 + \ldots + W_r}{\sigma\sqrt{rp}} \xrightarrow{w} N \sim \mathcal{N}(0,1).$$

Now in order to obtain the asymptotic normality of $\dfrac{V_1 + \ldots + V_r}{\sigma\sqrt{n}}$ it suffices to prove that Δ_n converges to zero in probability ([BI1]). To this aim we use

(1.46), we have

$$m_n = O\left(r\left(\frac{p^{1/2}}{\sqrt{n}/r}\right)^{\gamma/2\gamma+1} \alpha([n^{1/4}])^{\frac{2\gamma}{2\gamma+1}}\right)$$

$$= O\left((\log n)^{\frac{7\gamma+2}{4\gamma+2}} \alpha([n^{1/4}])^{\frac{2\gamma}{2\gamma+1}}\right)$$

so that (1.43) easily yields

$$\lim_{n\to\infty} m_n = 0.$$

Finally consider the identity

$$\frac{S_n}{\sigma\sqrt{n}} = \frac{\sum_1^r V_j}{\sigma\sqrt{n}} + \frac{\sum_1^r V_j'}{\sigma\sqrt{n}} + \frac{R_n}{\sigma\sqrt{n}}$$

where

$$R_n = X_{r(p+q)+1} + \ldots + X_n \quad \text{if } r(p+q) < n$$
$$= 0 \qquad\qquad\qquad\qquad \text{otherwise}$$

It remains to show that $\dfrac{\sum_1^r V_j'}{\sigma\sqrt{n}}$ and $\dfrac{R_n}{\sigma\sqrt{n}}$ converge to zero in probability.

First we clearly have

$$\frac{\sum_1^r V_j'}{\sigma\sqrt{qr}} \xrightarrow{W} N \sim \mathcal{N}(0,1)$$

therefore

$$\frac{\sum_1^r V_j'}{\sigma\sqrt{n}} = \sqrt{\frac{qr}{n}}\frac{\sum_1^r V_j'}{\sigma\sqrt{qr}} \xrightarrow{P} 0$$

since $\sqrt{\dfrac{qr}{n}} \sim \sqrt{\dfrac{\mathrm{Log} n}{n^{1/4}}}$.

Second, using Tchebychev's inequality we get

$$\frac{R_n}{\sigma\sqrt{n}} \xrightarrow{P} 0$$

Collecting the above results we obtain (1.44). ∎

Notice that Oodeira-Yoshihara in [OY] have shown a functional central limit theorem when the assumptions of Theorem 1.7 hold. We do not state it because we will not use it in the sequel.

Notes

The strong mixing condition has been introduced by ROSENBLATT ([RO]) in 1956. The basic properties of strong mixing conditions are studied by BRADLEY in [BR2] but the most complete reference should be the book by DOUKHAN ([DK] 1994). The coupling lemma's are from [BE] and [BR1] with a slight improvement due to RHOMARI ([RH] 1994). The optimal RIO's inequality is in ([RI] 1993).

The second part of Lemma 1.3 is given in [BO2]. Concerning the exponential inequalities (1.16) and (1.18) some improvements may be found in the BENNETT's paper [BE]. The original forms and the proof's method of Theorems 1.3 and 1.4 are obtained in [BO1]. The present statement is an amelioration using some ideas of RHOMARI. Related inequalities may be found in [DK] and [CA].

Theorem 1.5 is a result of DAVIDOV ([DA]), Theorem 1.6 is an easy consequence of the exponential inequalities and Theorem 1.7 was obtained by IBRAGIMOV in [IB], here the proof is simpler than the original one since we use the powerful Bradley's lemma.

Chapter 2

Density estimation for discrete time processes

This chapter deals with nonparametric density estimation for sequences of correlated random variables.

We consider here the popular convolution kernel estimate. That natural and simple method has well resisted to the other suggested estimates such as Projection estimates (in particular wavelets), Nearest Neighbour estimates, Recursive estimates and, more generally, estimates based on δ- sequences.

We shall see that, under mild conditions, it is possible to obtain the same convergence rates and the same asymptotic distribution as in the i.i.d. case.

The asymptotic behaviour of the kernel estimate in some non regular cases (errors in variables, chaotic data, singular distribution) is studied at the end of the chapter.

Let us define a d-dimensional *kernel* as an application $K : \mathbb{R}^d \longrightarrow \mathbb{R}$ where K is a bounded symmetric density with respect to Lebesgue measure, such that

$$\lim_{\|u\| \to \infty} \| u \|^d K(u) = 0$$

and

$$\int_{\mathbb{R}^d} \| u \|^2 K(u) du < +\infty.$$

Given a kernel K and a *smoothing parameter* h we set

(2.1) $$K_h(u) = \frac{1}{h^d} K\left(\frac{u}{h}\right) \quad , \quad u \in \mathbb{R}^d .$$

Typical examples of kernels are :

- The naive kernel $K = 1_{[-\frac{1}{2},+\frac{1}{2}]^d}$

- The normal kernel $K(u) = (2\pi)^{-d/2} \exp(-\parallel u \parallel^2 /2)$, $u \in \mathbb{R}^d$

- The Epanechnikov kernel $K(u) = \left(\frac{3}{4}\sqrt{5}\right)^d \prod_{i=1}^{d} \left(1 - \frac{u_i^2}{5}\right) 1_{[-\sqrt{5},+\sqrt{5}]}(u_i)$,
  $u = (u_1, \ldots, u_d) \in \mathbb{R}^d$.

2.1   Density estimation

Let $(X_t, t \in \mathbb{Z})$ be a $\mathbb{R}^d$-valued stochastic process. Suppose that the X_t's have a common density f and $X_1, \ldots, X_n$ are observed.

An estimate for f cannot be constructed directly from the *empirical measure*

$$(2.2) \qquad \mu_n = \frac{1}{n} \sum_{t=1}^{n} \delta_{(X_t)}$$

since μ_n is not absolutely continuous with respect to Lebesgue measure over $\mathbb{R}^d$. So in order to obtain such an estimate it is necessary to transform μ_n in a suitable way. The kernel method consists in a regularization of μ_n by convolution with a smoothed kernel, leading to the *kernel estimator*

$$(2.3) \qquad f_n(x) = (\mu_n * K_{h_n})(x) , \quad x \in \mathbb{R}^d ,$$

which can be written

$$(2.4) \qquad f_n(x) = \frac{1}{nh_n^d} \sum_{t=1}^{n} K\left(\frac{x - X_t}{h_n}\right) , \quad x \in \mathbb{R}^d .$$

Practical considerations leading to f_n are discussed in S.2.

Clearly the choice of (h_n) is crucial for the efficiency of f_n. In fact, it appears that, under some general assumptions, the conditions

$$(2.5) \qquad h_n \to 0(+) , \quad nh_n^d \to +\infty \quad (n \to +\infty)$$

are necessary and sufficient for the consistency of f_n. This will be clarified below; from now on, we do suppose that (2.5) is satisfied unless otherwise stated.

2.2 Optimal asymptotic quadratic error

Let us begin the study of (f_n) by evaluating of the asymptotic quadratic error. We will show that, under mild conditions, this error turns out to be the same as in the i.i.d. case.

We need some notations and assumptions : we suppose that for each couple (t, t'), $t \neq t'$ the random vector $(X_t, X_{t'})$ has a density and we set

$$g_{t,t'} = f_{(X_t, X_{t'})} - f \otimes f \quad , \quad t \neq t' .$$

Furthermore we suppose that $g_{t,t'}$ satisfies one of these two hypothesis :

- H_1 . $\delta_p = \sup_{|t'-t| \geq 1} \| g_{t,t'} \|_p < +\infty$ for some $p \in]2, +\infty]$.

- H_2 . $|g_{t,t'}(z') - g_{t,t'}(z)| \leq \ell \| z' - z \| \; ; \; z, z' \in \mathbb{R}^{2d}$ for some constant ℓ.

On the other hand let us denote by $\mathcal{C}_{2,d}(b)$ the space of twice continuously differentiable real valued functions f, defined on $\mathbb{R}^d$, and such that $\| f \|_\infty \leq b$ and $\| f^{(2)} \|_\infty \leq b$ where $f^{(2)}$ denotes any partial derivative of order 2 for f. We suppose that the density f belongs to $\mathcal{C}_{2,d}(b)$.

Finally (X_t) is supposed to be 2-α-*mixing* (see 1.4) and such that

$$(2.6) \qquad \alpha^{(2)}(k) \leq \gamma k^{-\beta} \quad , \quad k \geq 1$$

for some positive constants γ and β.

Now we state the result.

THEOREM 2.1 *If $f(x) > 0$, if* H_1 *(resp.* H_2*) holds and if $\beta > 2\dfrac{p-1}{p-2}$ (resp. $\beta > \dfrac{2d+1}{d+1}$) then the choice $h_n = c_n n^{-1/(d+4)}$ where $c_n \to c > 0$ leads to*

$$(2.7) \qquad n^{4/(d+4)} E[f_n(x) - f(x)]^2 \to C(c, K, f) > 0$$

where

$$C(c, K, f) = \frac{c^4}{4} \left(\sum_{1 \leq i,j \leq d} \frac{\partial^2 f}{\partial x_i \partial x_j}(x) \int u_i u_j K(u) du \right)^2 + \frac{f(x)}{c^d} \int K^2 .$$

Proof

The following decomposition is valid :

$$
\begin{aligned}
E(f_n(x) - f(x))^2 &= (Ef_n(x) - f(x))^2 + \frac{1}{n}\mathrm{Var}K_{h_n}(x - X_1) \\
&\quad + \frac{1}{n(n-1)} \sum_{1 \leq |t'-t| \leq n-1} \mathrm{Cov}\left(K_{h_n}(x - X_t), K_{h_n}(x - X_{t'})\right) \\
&= : B_n^2(x) + \widetilde{V}f_n(x) + C_n \ .
\end{aligned}
$$

(2.8)

We treat each term separately. First we consider the bias :

$$
\begin{aligned}
B_n(x) &= \int_{\mathbb{R}^d} K_{h_n}(x - u)[f(u) - f(x)]du \\
&= \int_{\mathbb{R}^d} K(v)[f(x - h_n v) - f(x)]dv \ .
\end{aligned}
$$

By using Taylor's formula and the symmetry of K we get

$$
B_n(x) = \frac{h_n^2}{2} \int \sum_{1 \leq i,j \leq d} \frac{\partial^2 f}{\partial x_i \partial x_j}(x - \theta h_n v)v_i v_j K(v)dv
$$

where $0 < \theta < 1$. Thus a simple application of Lebesgue dominated convergence theorem gives

$$
(2.9) \qquad h_n^{-4}B_n^2(x) \to \frac{1}{4}\left(\sum_{1 \leq i,j \leq d} \frac{\partial^2 f}{\partial x_i \partial x_j}(x) \int v_i v_j K(v)dv\right)^2 .
$$

Now $\widetilde{V}f_n(x)$ is nothing else but the variance of f_n in the i.i.d. case. It can be written

$$
\widetilde{V}f_n(x) = \frac{1}{n}\left[\int K_{h_n}^2(x - u)f(u)du - \left(\int K_{h_n}(x - u)f(u)du\right)^2\right] ,
$$

then writing $\mathbb{R}^d = \{u :\| u \| \leq \eta\} \cup \{u :\| u \| > \eta\}$ where η is small enough it is easy to infer that

$$
(2.10) \qquad h_n^d \int K_{h_n}^2(x - u)f(u)du \to f(x)\int K^2
$$

and

$$
(2.11) \qquad \int K_{h_n}(x - u)f(u)du \to f(x).
$$

(In fact (2.10) and (2.11) are two forms of a famous **Bochner's lemma** (see [PA]) and [CA]-[LE] in Chapter 4).

Hence (2.10) and (2.11) imply

$$(2.12) \qquad nh_n^d \tilde{V} f_n(x) \to f(x) \int K^2.$$

The covariance term C_n remains to be studied. First note that

$$\text{Cov}\,(K_{h_n}(x - X_t), K_{h_n}(x - X_{t'})) \;=\; \int K_{h_n}(x - u)K_{h_n}(x - v)g_{t,t'}(u, v)dudv$$
$$=\; :c_{t,t'}\,.$$

Thus, if H_1 holds, Hölder inequality yields

$$(2.13) \qquad |c_{t,t'}| \le \| g_{t,t'} \|_p\, h_n^{-2d/p}\, \| K \|_q^2$$

where $\dfrac{1}{p} + \dfrac{1}{q} = 1$.

On the other hand Billingsley's inequality (1.11) entails

$$(2.14) \qquad |c_{t,t'}| \le 4h_n^{-2d}\, \| K \|_\infty^2\, \alpha^{(2)}(|t' - t|)\,,$$

thus
$$|c_{t,t'}| \le \gamma_n(|t' - t|)$$
where

$$\gamma_n(|t' - t|) = \max\left(\delta_p h_n^{-2d/p}\, \| K \|_q^2, 4h_n^{-2d}\, \| K \|_\infty^2\, \alpha^{(2)}(|t' - t|)\right)\,.$$

Consequently

$$|C_n| \le \frac{2}{n}\sum_{t=1}^{n-1}\gamma_n(t)$$

which implies

$$|C_n| \le \frac{2}{n}\left[\sum_{t=1}^{u_n}\delta_p\, \| K \|_q^2\, h_n^{-2d/p} + h_n^{-2d}\sum_{t>u_n} 4\, \| K \|_\infty^2\, \gamma t^{-\beta}\right]$$

where $u_n \sim h_n^{-2d/q\beta}$.

Now elementary calculations give

$$(2.15) \qquad nh_n^d|C_n| = o(1).$$

Finally using the decomposition (2.8), the asymptotic results (2.9), (2.12), (2.15) and the fact that $h_n \cong cn^{-1/(4+d)}$ we obtain the desired result (2.7).

When H_2 holds the proof is similar. The only difference lies in the overestimation of $c_{t,t'}$: using (2.14) in Lemma 1.3 we get

$$(2.16) \qquad |c_{t,t'}| \leq \gamma(d,\ell) \left[\alpha^{(2)}(|t'-t|)\right]^{1/(2d+1)}$$

and consequently

$$|C_n| \leq \frac{2}{n} \left[\sum_{t=1}^{v_n} \gamma(d,\ell) \left[\alpha^{(2)}(|t'-t|)\right]^{1/(2d+1)} + h_n^{-2d} \sum_{t>v_n} 4 \parallel K \parallel_\infty^2 \gamma t^{-\beta}\right]$$

then the choice $v_n \sim h_n^{-(2d+1)/\beta}$ leads to $nh_n^d|C_n| = o(1)$ since $\beta > \dfrac{2d+1}{d+1}$. $\blacksquare$

In order to obtain a **uniform result** let us introduce the family $\mathcal{X} = \mathcal{X}(\ell, b, \gamma, \beta)$ of $\mathbb{R}^d$-valued stochastic processes $(X_t, t \in \mathbb{Z})$ satisfying H_2 for a fixed ℓ, such that $f \in \mathcal{C}_{2,d}(b)$ and satisfying (2.6) with the same γ and $\beta > \dfrac{2d+1}{d+1}$. Then we clearly have

COROLLARY 2.1
$\sup_{X \in \mathcal{X}} \overline{\lim}_{n \to \infty} n^{4/(d+4)} \sup_{x \in \mathbb{R}^d} E(f_n(x) - f(x))^2 < +\infty$.

Finally it can be proved that $n^{-4/(d+4)}$ is the **best attainable rate in a minimax sense**. We shall establish this kind of result in a more general context in Chapter 4 (see Theorem 4.3).

2.3   Uniform almost sure convergence

The quadratic error is a useful measure of the accuracy of a density estimate. However it is not completely satisfactory since it does not provide information concerning the shape of the graph of f whereas the similarity between the graph of f_n and that of f is crucial for the user.

A good measure of this similarity should be the uniform distance between f_n and f. In the current section we study the magnitude of this distance.

Let us introduce the notion of "geometrically strongly mixing" (GSM) process. We will say that (X_t) is GSM if there exist $c_0 > 0$ and $\rho \in [0, 1[$ such that
$$(2.17) \qquad \alpha(k) \leq c_0 \rho^k \quad k \geq 1.$$

Note that usual linear processes are GSM (see 1.3).

The following lemma deals with simple almost sure convergence of f_n for a GSM process.

LEMMA 2.1 *Let $(X_t, t \in \mathbb{Z})$ be a strictly stationary GSM $\mathbb{R}^d$-valued process and let f be the density of X_t.*

1) *If f is continuous at x and if $\dfrac{nh_n^d}{(\mathrm{Log}\,n)^2} \to +\infty$ then*

$$(2.18) \qquad f_n(x) \to f(x) \quad a.s.$$

2) *If $f \in C_{2,d}(b)$ for some b and if $h_n = c_n \left(\dfrac{\mathrm{Log}\,n}{n}\right)^{\frac{1}{d+4}}$ where $c_n \to c > 0$, then for all $x \in \mathbb{R}^d$ and all integer k*

$$(2.19) \qquad \frac{1}{\mathrm{Log}_k n} \left(\frac{n}{\mathrm{Log}\,n}\right)^{\frac{2}{d+4}} (f_n(x) - f(x)) \to 0 \quad a.s. .$$

Proof

1) The continuity of f at x and (2.11) yield

$$E f_n(x) \to f(x) ,$$

thus it suffices to prove that

$$f_n(x) - E f_n(x) \to 0 \quad \text{a.s. .}$$

For that purpose we apply inequality (1.26) to the random variables

$$Y_{tn} = K_{h_n}(x - X_t) - E K_{h_n}(x - X_t) \ , 1 \le t \le n.$$

Note that $|Y_{tn}| \le \| K \|_\infty h_n^{-d}$ an choose $q = q_n = \left[\sqrt{n}h_n^{-d/2}\right]$. Then by using the GSM assumption (2.17) and BILLINGSLEY's inequality (1.11) it is easy to infer that

$$\sigma^2(q) = O(ph_n^{-d}) ,$$

therefore for all $\varepsilon > 0$ we have for n large enough

$$v^2(q) \le \| K \|_\infty h_n^{-d}\varepsilon.$$

Hence

$$\begin{aligned}
(2.20) \qquad P(|f_n(x) - E f_n(x)| \ &> \ \varepsilon) \le 4\exp\left(-\frac{\varepsilon^2}{8 \| K \|_\infty}qh_n^d\right) \\
&+ \ 22\left(1 + \frac{4 \| K \|_\infty h_n^{-d}}{\varepsilon}\right)^{1/2} qc_0\rho^{\left[\frac{n}{2q}\right]}
\end{aligned}$$

which implies

$$P(|f_n(x) - Ef_n(x)| > \varepsilon) \le \beta \exp\left(-\gamma\sqrt{n}h_n^{d/2}\right)$$

where $\beta = \beta(\varepsilon, K, d)$ and $\gamma = \gamma(\varepsilon, K, d)$ are strictly positive.

Now setting $u_n = \dfrac{nh_n^d}{(\mathrm{Log}n)^2}$ we obtain the bound

(2.21) $$P(|f_n(x) - Ef_n(x)| > \varepsilon) \le \frac{\beta}{n^{\gamma\sqrt{u_n}}}$$

thus

$$\sum_n P(|f_n(x) - Ef_n(x)| > \varepsilon) < +\infty \quad , \quad \varepsilon > 0$$

and the Borel-Cantelli lemma entails the desired result. ∎

2) Concerning the bias we have established the following expression in the proof of Theorem 2.1

$$Ef_n(x) - f(x) = \frac{h_n^2}{2} \int \sum_{1 \le i,j \le d} \frac{\partial^2 f}{\partial x_i \partial x_j}(x - \theta h_n v)v_i v_j K(v)dv$$

where $\theta = \theta(f, x, h_n, v) \in \,]0, 1[$.

Then since $f \in C_{2,d}(b)$ we have

(2.22) $$|Ef_n(x) - f(x)| \le ah_n^2$$

where $a = a(d, K, b)$ does not depend on n.

Now set

$$\varepsilon_n = \mathrm{Log}_k n \left(\frac{\mathrm{Log}n}{n}\right)^{\frac{2}{d+4}} \quad ,$$

we get

$$\varepsilon_n^{-1}|Ef_n(x) - f(x)| \le \frac{ac_n^2}{\mathrm{Log}_k n}$$

and the bound vanishes at infinity. Thus we only need to show that

$$\varepsilon_n^{-1}|f_n(x) - Ef_n(x)| \to 0 \quad \text{a.s.}$$

To this aim we again apply inequality (1.26) with

$$q = \left[(\mathrm{Log}_k n)^2(\mathrm{Log}n)^{\frac{2}{4+d}} \; n^{\frac{2+d}{4+d}}\right] \;.$$

Substituting in (1.26) and symplifying we obtain for n large enough

$$(2.23) \qquad P\left(\varepsilon_n^{-1}|f_n(x) - Ef_n(x)| > \eta\right) \le v_n , \quad \eta > 0$$

where

$$v_n = 4\exp\left(-c_1 \text{Log}n(\text{Log}_k n)^2\right) + c_2 \exp\left(-c_3 \frac{n^{2/(4+d)}}{(\text{Log}_k n)^2 (\text{Log}n)^{2/(4+d)}}\right)$$

and where the c_i's are strictly positive constants.

Therefore for all $\eta > 0$ we have

$$\sum_n P\left(\varepsilon_n^{-1}|f_n(x) - Ef_n(x)| > \eta\right) < +\infty$$

hence (2.19) by using Borel-Cantelli lemma. ∎

Now we can state a uniform result on an increasing sequence of compact sets.

THEOREM 2.2 *Let $(X_t, t \in \mathbb{Z})$ be a strictly stationary GSM $\mathbb{R}^d$-valued process and let f be the density of X_t.*
Let f_n be the kernel estimate associated with a kernel K which satisfies a Lipschitz condition.

1) If f is uniformly continuous and if $\dfrac{nh_n^d}{(\text{Log}n)^2} \to +\infty$ then, for all $\gamma > 0$,

$$(2.24) \qquad \sup_{\|x\| \le n^\gamma} |f_n(x) - f(x)| \to 0 \quad a.s..$$

2) If $f \in C_{2,d}(b)$ for some b and if $h_n = c_n \left(\dfrac{\text{Log}n}{n}\right)^{\frac{1}{d+4}}$ where $c_n \to c > 0$
then, for all $\gamma > 0$ and all integer k

$$(2.25) \qquad \frac{1}{\text{Log}_k n} \left(\frac{n}{\text{Log}n}\right)^{\frac{2}{d+4}} \sup_{\|x\| \le n^\gamma} |f_n(x) - f(x)| \to 0 \quad a.s. .$$

Proof

1) f being a uniformly continuous integrable function, it is therefore bounded. Thus it is easy to see that for all $\delta > 0$

$$\sup_{x \in \mathbb{R}^d} |Ef_n(x) - f(x)| \le \sup_{x \in \mathbb{R}^d} \sup_{\|y\| \le \delta} |f(x - y) - f(x)|$$
$$(2.26)$$
$$+ \; 2 \, \| f \|_\infty \int_{\|z\| > \delta h_n^{-1}} K(z)dz$$

Choosing δ^{-1} and n large enough that bound can be made arbitrarily small, hence

$$\sup_{x \in \mathbb{R}^d} |Ef_n(x) - f(x)| \to 0 \ .$$

Now we have to show that

$$\sup_{\|x\| \leq n^\gamma} |f_n(x) - Ef_n(x)| \to 0 \quad \text{a.s.} \ , \quad \gamma > 0$$

where for convenience we take $\| \ . \ \|$ as the sup norm on $\mathbb{R}^d$, defined by $\| (x_1, \ldots, x_d) \| = \sup_{1 \leq i \leq d} |x_i|$. In the sequel we may and do suppose that $\gamma > 1$.

The idea of the proof is to consider a covering of

$$B_n = \{x : \| x \| \leq n^\gamma\}$$

by ν_n^d closed hypercubes $B_{jn} = \left\{ x : \| x - x_{jn} \| \leq \dfrac{n^\gamma}{\nu_n} \right\}$, $1 \leq j \leq \nu_n^d$ such that $\overset{\circ}{B}_{jn} \cap \overset{\circ}{B}_{j'n} = \emptyset$, $1 \leq j, j' \leq \nu_n^d$, $j \neq j'$ and to write

$$\sup_{\|x\| \leq n^\gamma} |f_n(x) - Ef_n(x)| = \sup_{1 \leq j \leq \nu_n^d} \Delta_{jn} =: \Delta_n$$

where $\Delta_{jn} = \sup_{x \in B_{jn}} |f_n(x) - Ef_n(x)|$.

Now by assumption there exists $\ell > 0$ such that

$$|K(u') - K(u)| \leq \ell \| u' - u \| \ ; \ u, u' \in \mathbb{R}^d \ .$$

Consequently

$$(2.27) \qquad |f_n(x) - f_n(x_{jn})| \leq \frac{\ell}{h_n^{d+1}} \frac{n^\gamma}{\nu_n} \ , \ x \in B_{jn} \ , \ 1 \leq j \leq \nu_n^d,$$

and similarly

$$(2.28) \qquad |Ef_n(x) - Ef_n(x_{jn})| \leq \frac{\ell}{h_n^{d+1}} \frac{n^\gamma}{\nu_n} \ , \ x \in B_{jn} \ , \ 1 \leq j \leq \nu_n^d$$

then, choosing $\nu_n = \left[2\ell n^\gamma h_n^{d+1} \mathrm{Log}_2 n \right] + 1$

it follows that
$$\Delta_n \leq \Delta'_n + \frac{1}{\text{Log}_2 n}$$
where $\Delta'_n = \sup_{1 \leq j \leq \nu_n^d} |f_n(x_{jn}) - Ef_n(x_{jn})|$.
Now, for all $\varepsilon > 0$

$$P(\Delta'_n > \varepsilon) \leq \sum_{j=1}^{\nu_n^d} P(|f_n(x_{jn}) - Ef_n(x_{jn})| > \varepsilon)$$

and noting that (2.21) does not depend on x we get
$$P(\Delta'_n > \varepsilon) \leq \nu_n^d \frac{\beta}{n^\gamma \sqrt{u_n}}$$
where $u_n \to +\infty$, hence $\Sigma P(\Delta'_n > \varepsilon) < +\infty$ which implies $\Delta'_n \to 0$ a.s.
which in turn implies $\Delta_n \to 0$ a.s. and the proof of (2.24) is complete. ■

2) Since $f \in C_{2,d}(b)$ we may apply (2.22) and setting again
$$\varepsilon_n = \text{Log}_k n \left(\frac{\text{Log} n}{n}\right)^{\frac{2}{d+4}}$$
we obtain

$$\varepsilon_n^{-1} \sup_{\|x\| \leq n^\gamma} |Ef_n(x) - f(x)| \leq \frac{ac_n^2}{\text{Log}_k n}$$

which shows the uniform convergence of the bias.

It remains to be shown that

$$\varepsilon_n^{-1} \sup_{\|x\| \leq n^\gamma} |f_n(x) - Ef_n(x)| \longrightarrow 0 \quad \text{a.s..}$$

For that purpose we again consider a covering of B_n by hypercubes B_{jn}, $1 \leq j \leq \nu_n^d$ but this time we choose $\nu_n = \left[n^{\gamma + \frac{3}{d+4}}\right]$.

Then by using (2.27) and (2.28) we obtain

$$\varepsilon_n^{-1}\Delta_n \leq \varepsilon_n^{-1}\Delta'_n + \frac{2\ell}{\text{Log}_k n} \frac{1}{(\text{Log} n)^{3/(d+4)}} \;\cdot$$

Now (2.23) entails

$$P\left(\varepsilon_n^{-1}\Delta'_n > \eta\right) \leq \nu_n^d v_n \quad, \quad \eta > 0 \;.$$

Clearly $\Sigma \nu_n^d v_n < +\infty$ and (2.25) follows. ■

It is noteworthy that the obtained rate in (2.25) is optimal. In fact by applying a theorem of STUTE (see [ST]) for i.i.d. X_t's we obtain for all $A > 0$ and $f > 0$

$$\left(\frac{n}{\text{Log} n}\right)^{\frac{2}{d+4}} \sup_{\|x\| \le A} \left|\frac{f_n(x) - f(x)}{f(x)}\right| \longrightarrow \left(\frac{d}{(d+4)c^d} \int K^2\right)^{1/2} \quad \text{a.s.} \ .$$

We now study the uniform behaviour of f_n over the entire space. The results are summarized in the following corollary.

COROLLARY 2.2 *Let f_n be the estimator associated with the normal kernel. Then*

1) *If assumptions of Theorem 2.1 hold and if in addition $E \parallel X_0 \parallel < \infty$ we have*

(2.29) $$\sup_{x \in \mathbb{R}^d} |f_n(x) - f(x)| \longrightarrow 0 \quad \text{a.s.} \ .$$

2) *If assumptions of Theorem 2.2 hold and if*

(2.30) $$\overline{\lim}_{\|x\| \to \infty} \parallel x \parallel^{d+2} f(x) < +\infty$$

then for every integer k

(2.31) $$\frac{1}{\text{Log}_k n} \left(\frac{n}{\text{Log} n}\right)^{\frac{2}{d+4}} \sup_{x \in \mathbb{R}^d} |f_n(x) - f(x)| \longrightarrow 0 \quad \text{a.s.} \ .$$

It may be shown that (2.29) and (2.31) are still valid if K has compact support.

Proof

1) Since (2.24) is valid it suffices to establish that

$$\sup_{\|x\| > n^\gamma} |f_n(x) - f(x)| \longrightarrow 0 \quad \text{a.s..}$$

Now f is an integrable uniformly continuous function, hence

(2.32) $$\sup_{\|x\| > n^\gamma} f(x) \longrightarrow 0$$

and it remains to check that

(2.33) $$\sup_{\|x\| > n^\gamma} f_n(x) \longrightarrow 0 \quad \text{a.s.} \ .$$

First note that we may choose $\gamma > 2$, then, by Markov inequality

$$P\left(\bigcup_{t=1}^n \parallel X_t \parallel > \frac{n^\gamma}{2}\right) \le \sum_1^n P\left(\parallel X_t \parallel > \frac{n^\gamma}{2}\right) \le \frac{2E \parallel X_0 \parallel}{n^{\gamma-1}} \ .$$

Therefore by Borel-Cantelli lemma

$$P\left(\overline{\lim}\bigcup_{t=1}^{n}\parallel X_t\parallel>\frac{n^\gamma}{2}\right)=0\,.$$

Now let us set $\Omega_0=\underline{\lim}\left\{\bigcap_{t=1}^{n}\parallel X_t\parallel\le\frac{n^\gamma}{2}\right\}$.

For every $\omega\in\Omega_0$ there exists $n_0(\omega)$ such that $n\ge n_0(\omega)$ implies
$\parallel X_t(\omega)\parallel\le\frac{n^\gamma}{2}$; $t=1,\dots,n$. Therefore, if $\parallel x\parallel>n^\gamma$ we have
$\parallel\frac{x-X_t(\omega)}{h_n}\parallel>\frac{n^\gamma}{2h_n}$; $t=1,\dots,n$.

Then choosing $x_n\in\mathbb{R}^d$ such that $\parallel x_n\parallel=\frac{n^\gamma}{2h_n}$ we obtain

$$K\left(\frac{x-X_t(\omega)}{h_n}\right)<K(x_n)\quad t=1,\dots,n$$

hence

$$\sup_{\parallel x\parallel>n^\gamma}f_n(x,\omega)\le\frac{1}{nh_n^d}\left(\sum_{t=1}^{n_0(\omega)}K\left(\frac{x-X_t}{h_n}\right)+(n-n_0(\omega))K(x_n)\right)$$

and finally

(2.34) $$\sup_{\parallel x\parallel>n^\gamma}f_n(x,\omega)\le\frac{(2\pi)^{-d/2}}{nh_n^d}\left(n_0(\omega)+\exp\left(-\frac{1}{2}\frac{n^{2\gamma}}{4h_n^2}\right)\right),$$

$\omega\in\Omega_0,n\ge n_0(\omega)$, which implies (2.33). ∎

2) The proof is similar. From (2.30) we get

$$\varepsilon_n^{-1}\sup_{\parallel x\parallel>n^\gamma}f_n(x,\omega)\longrightarrow0\ ,\ \omega\in\Omega_0$$

and by using Theorem 2.2 we come to (2.31). ∎

2.4   Asymptotic normality

Let $(X_t,t\in\mathbb{Z})$ be a $\mathbb{R}^d$-valued Strictly Stationary Process with marginal distributions satistying the following conditions

$C_1)$ The density $f_{t_1,\dots,t_4}$ of $(X_{t_1},\dots,X_{t_4})$ exists whenever $t_1<t_2<t_3<t_4$ and

$$\sup_{t_1<t_2<t_3<t_4}\parallel f_{t_1,\dots,t_4}\parallel_\infty<+\infty\,.$$

C_2) $\sup_{t_1 < t_2} \| f_{t_1, t_2} - f \otimes f \|_\infty < +\infty$.

C_3) $f \in \mathcal{C}_{2,d}(b)$.

Then we have

THEOREM 2.3 *If C_1, C_2, C_3 hold, if $\alpha(k) = O(k^{-\beta})$ where $\beta \geq 2$ and if*
$h_n = \dfrac{c}{\text{LogLog} n} n^{-\frac{1}{d+4}}$, *$c > 0$ then for all integer m and all distincts x_i's such that $f(x_i) > 0$*

$$(2.35) \qquad \left(n h_n^d \right)^{1/2} \left(\frac{f_n(x_i) - f(x_i)}{(f_n(x_i) \int K^2)^{1/2}} \ , \ 1 \leq i \leq m \right) \xrightarrow{w} N^{(m)}$$

where $N^{(m)}$ denotes a random vector with standard normal distribution in $\mathbb{R}^m$.

Before the proof, let us mention that the precise form of (2.35) is useful for constructing confidence sets for $(f(x_1), \dots, f(x_m))$.

Proof
   We first show that

$$(2.36) \ \left(n h_n^d \right)^{1/2} \left(\frac{f_n(x_i) - E f_n(x_i)}{(f(x_i) \int K^2)^{1/2}}, 1 \leq i \leq m \right) \xrightarrow{w} \left(N_1^{(m)}, \dots, N_m^{(m)} \right) .$$

According to the Cramer-Wold device (see [BI] p.49) it suffices to prove that whatever $(\lambda_1, \dots, \lambda_m) \in \mathbb{R}^m$ such that $\displaystyle\sum_{i=1}^m |\lambda_i| \neq 0$

$$(2.37) \qquad \left(n h_n^d \right)^{1/2} \sum_{i=1}^m \lambda_i \frac{f_n(x_i) - E f_n(x_i)}{(f(x_i) \int K^2)^{1/2}} \xrightarrow{w} \sum_{i=1}^m \lambda_i N_i^{(m)} .$$

Let us set

$$S_n = \sum_{t=1}^n Y_{t,n}$$

where

$$Y_{t,n} = \sum_{i=1}^m \frac{\lambda_i}{(f(x_i) \int K^2)^{1/2}} \left[K \left(\frac{x_i - X_t}{h_n} \right) - E K \left(\frac{x_i - X_t}{h_n} \right) \right] \ ;$$

$t = 1, \dots, n$, and consider the blocks
$V_{1n} = Y_{1n} + \dots + Y_{pn}$, $V'_{1n} = Y_{p+1,n} + \dots + Y_{p+q,n}$,...,
$V_{rn} = Y_{(r-1)(p+q)+1,n} + \dots + Y_{rp+(r-1)q,n}$, $V'_{rn} = Y_{rp+(r-1)q+1,n} + \dots + Y_{r(p+q),n}$,
where

$$r(p+q) \leq n < r(p+q+1) .$$

From Bradley's lemma 1.2 there exist i.i.d. random variables $W_{1n}, \ldots,$ W_{rn} such that $P_{W_{jn}} = P_{V_{jn}}$ and

$$P(|V_{jn} - W_{jn}| > \xi_n) \leq 11 \left(\frac{\| V_{jn} + c_n \|_\infty}{\xi_n} \right)^{1/2} \alpha(q) \; ; j = 1, \ldots, r \;,$$

with $c_n = 2p \sum_{i=1}^{m} \frac{|\lambda_i| \| K \|_\infty}{(f(x_i) \int K^2)^{1/2}}$, $\xi_n = \varepsilon \left(rph_n^d \right)^{1/2}, \varepsilon > 0.$

Therefore, if r, p, q tend to infinity we have

(2.38) $$P(|V_{jn} - W_{jn}| > \xi_n) = O \left(\frac{p^{1/4}\alpha(q)}{r^{1/4}h_n^{d/4}} \right) \;.$$

Now consider

(2.39) $$\Delta_n = \frac{\sum_{j=1}^{r} V_{jn}}{(rph_n^d)^{1/2}} - \frac{\sum_{j=1}^{r} W_{jn}}{(rph_n^d)^{1/2}} \;,$$

by using (2.38) we get for all $\varepsilon > 0$

$$P \left(|\Delta_n| > \varepsilon \right) = O(p^{1/4}r^{3/4}\alpha(q)h_n^{-d/4}) \;.$$

Then choosing $r \sim n^a$, $p \sim n^{1-a}$, $q \sim n^c$, $0 < a < 1, 0 < c < 1$ where a and c will be specified below, we obtain

(2.40) $$P(|\Delta_n| > \varepsilon) = O \left(n^{1/2\left(a - 2c\beta + \frac{d+2}{d+4}\right)} (\mathrm{LogLog} n)^{d/4} \right)$$

which tends to zero provided that $c > \frac{1}{2\beta} \left(a + \frac{d+2}{d+4} \right)$, thus

(2.41) $$\Delta_n \to 0 \quad \text{in probability} \;.$$

We now prove the asymptotic normality of

$$Z_n = \frac{\sum_{j=1}^{r} W_{jn}}{(rph_n^d)^{1/2}} \;, \; n \geq 1 \;.$$

To this aim we apply the Liapounov condition (see [BI] p. 44). It suffices to show that

(2.42) $$z_n = \frac{\sum_{j=1}^{r} E|W_{jn}|^3}{(r\mathrm{Var}W_{1n})^{3/2}} \to 0 \;.$$

By using the same arguments as in Theorem 2.1 it can be shown that

$$\mathrm{Var}W_{1n} = \mathrm{Var}V_{1n} \sim ph_n^d \sum_{i=1}^{m} \lambda_i^2 \ .$$

On the other hand, the uniform boundedness of $f_{t_1,\dots,t_4}$ implies

$$EW_{jn}^4 = O\left(p^4 h_n^{4d}\right) \ .$$

Now applying Schwarz inequality in (2.42) and the here above results we obtain

$$z_n = O\left(\frac{p}{r^{1/2}} h_n^d\right)$$

that is

$$z_n = O\left(n^{1-\frac{3a}{2}-\frac{d}{4+d}}(\mathrm{LogLog}n)^{-d}\right)$$

which tends to zero provided that $a \geq \dfrac{8}{3(d+4)}$.

Now write

$$S_n = \sum_{j=1}^{n} V_{jn} + \sum_{j=1}^{n} V_{jn}' + \delta_n \ ,$$

it is easy to see that

$$\mathrm{Var}\left(\frac{\displaystyle\sum_{j=1}^{r} V_{jn}'}{(rph_n^d)^{1/2}}\right) = O(n^{c-1+a})$$

and

$$\mathrm{Var}\left(\frac{\delta_n}{(rph_n^d)^{1/2}}\right) = O(n^{a-1})$$

which both tend to zero provided that $0 < c < 1 - a$.

These conditions are satisfied if $a < \dfrac{2\beta}{2\beta+1} - \dfrac{1}{2\beta+1}\dfrac{d+2}{d+4}$ which is compatible with $a \geq \dfrac{8}{3(d+4)}$ and $c > \dfrac{1}{2\beta}\left(a + \dfrac{d+2}{d+4}\right)$ since $\beta \geq 2$.

Collecting the above results we obtain

$$Z_n \xrightarrow{w} \sum_{i=1}^{m} \lambda_i N_i^{(m)}$$

$$\Delta_n \xrightarrow{p} 0$$

$$\left(rph_n^d\right)^{-1/2} \sum_{j=1}^{r} V'_{jn} \xrightarrow{p} 0$$

$$\left(rph_n^d\right)^{-1/2} \delta_n \xrightarrow{p} 0$$

which put together imply (2.36) by using [BI] (Theorem 4.1 p. 25).

Now in (2.36) $Ef_n(x_i)$ may be replaced by $f(x_i)$ since $f \in \mathcal{C}_{2,d}(b)$, in fact, using (2.9), we get

$$
\begin{aligned}
\left(nh_n^d\right)^{1/2} |Ef_n(x_i) - f(x_i)| &= \left(nh_n^d\right)^{1/2} O\left(h_n^2\right) \\
&= O\left((\mathrm{LogLog}n)^{-\frac{4+d}{2}}\right)
\end{aligned}
$$

Finally since Lemma 2.1 shows that

$$f_n(x_i) \xrightarrow{a.s.} f(x_i) \quad \imath = 1, \ldots, m,$$

in (2.36) we may also replace $f(x_i)$ by $f_n(x_i)$ and the proof of Theorem 2.3 is therefore complete. ∎

2.5   Non regular cases

In this section we will show that f_n may have an interesting behaviour in some unusual situations.

2.5.1   Chaotic data

We begin with an example which shows that asymptotic independence is not a necessary condition for obtaining sharp rates of convergence for f_n.

Let r be a mapping from E to E where E belongs to $\mathcal{B}_{\mathbb{R}^d}$ and $(X_t \, ; \, t = 0, 1, 2, \ldots)$ a sequence of *r.v.*'s which satisfies

$$(2.43) \qquad\qquad X_t = r(X_{t-1}) \quad ; \, t = 1, 2, \ldots$$

we suppose that the following conditions hold

$D_1)$ r preserves μ, a probability on E having a density f with respect to Lebesgue measure.

D_2) There exists $\gamma_0 > 0$, $\rho \in]0,1[$ and $c > 0$ such that

$$|\mu(B \cap r^{-t}(B)) - (\mu(B))^2| \le c\rho^t \ , \ \ t \ge 1$$

for each hypercube B in $\mathbb{R}^d$ satisfying $\mu(B) < \gamma_0$.

D_3) X_0 has a density g_0 and, for each $t \ge 1$, X_t has a density $\pi^t g_0$ such that

$$\sum_{t=1}^{+\infty} \| \pi^t g_0 - f \|_\infty < +\infty .$$

A typical example of a model satisfying D_1, D_2, D_3 should be the **p - adic Process**.

(2.44) $X_t = pX_{t-1} \, (\mathrm{mod}\, 1) \ , \ \ t \ge 1 \ ,$

where $p \ge 2$ is an integer and where X_0 has a density g_0 concentrated on $[0,1]$ and with a bounded derivative.
Other examples may be found in [LM] and [RU].

Clearly the process defined by (2.43) is not strongly mixing. However we have the following astonishing result.

THEOREM 2.4 *Let f_n^0 be the kernel estimate associated with the naive kernel $K = 1_{[-\frac{1}{2},+\frac{1}{2}]^d}$. Suppose that D_1, D_2, D_3 are satisfied and that $f \in C_{2,d}(b)$. Then if $h_n = c \left(\dfrac{\mathrm{Log}\, n}{n} \right)^{1/(d+4)}$, $c > 0$ we have*

(2.45) $E \left(f_n^0(x) - f(x) \right)^2 = O \left(\left(\dfrac{\mathrm{Log}\, n}{n} \right)^{4/d+4} \right) \ , \ x \in \overset{\circ}{E} .$

Thus despite the high dependence between the X_t's, the convergence rate of f_n^0 is the same as in the i.i.d. case, up to a logarithm. This fact comes from the chaotic character of r. Note that (2.45) shows that f_n^0 is useful for the empirical determination of the invariant measure of a dynamical system.

Proof of Theorem 2.4

First assume that $g_0 = f$, then from D_1, $\pi^t g_0 = f$ for each t and with similar notations as in Theorem 2.1 we have

$$E \left(f_n^0(x) - f(x) \right)^2 = B_n^2(x) + \widetilde{V} f_n(x) + C_n(x) ,$$

where B_n, $\widetilde{V} f_n$ and C_n are defined in (2.8).

Therefore by using (2.9), (2.12) and $h_n = c \left(\dfrac{\mathrm{Log}n}{n} \right)^{1/(d+4)}$ we obtain

(2.46) $$B_n^2(x) + \widetilde{V} f_n(x) = O\left(\left(\frac{\mathrm{Log}n}{n} \right)^{4/(d+4)} \right) .$$

Now setting $B_n = \left[-\dfrac{h_n}{2}, +\dfrac{h_n}{2} \right]^d$ we get

$$\begin{aligned} |P(X_t \in B_n, X_{t'} \in B_n) - P(X_t \in B_n)P(X_{t'} \in B_n)| &\leq 2\mu(B_n) \\ &\leq 2 \parallel f \parallel_\infty h_n^d . \end{aligned}$$

Thus from D_2 it follows that for n large enough

(2.47) $$\begin{aligned} |P(X_t \in B_n, X_{t'} \in B_n) &- P(X_t \in B_n)P(X_{t'} \in B_n)| \\ &\leq \varepsilon_n(|t' - t|) \end{aligned}$$

where

$$\varepsilon_n(|t' - t|) = \min(2 \parallel f \parallel_\infty h_n^d , c\rho^{|t'-t|}) .$$

From (2.47) we deduce that

$$|C_n(x)| \leq \frac{2}{n h_n^{2d}} \sum_{t=1}^{n-1} \varepsilon_n(t) .$$

Now choosing $w_n = \left[\dfrac{\mathrm{Log}h_n^{-d}}{\mathrm{Log}\rho^{-1}} \right]$ we find that

$$|C_n(x)| \leq \frac{4 \parallel f \parallel_\infty}{n h_n^d} \frac{\mathrm{Log}h_n^{-d}}{\mathrm{Log}\rho^{-1}} + \frac{2c}{n h_n^{2d}} \sum_{t>w_n} \rho^t$$

which implies

(2.48) $$|C_n(x)| = O\left(\left(\frac{\mathrm{Log}n}{n} \right)^{4/(d+4)} \right) ,$$

and (2.45) is proved if $g_0 = f$.

We now turn to the general case. First we have (with clear notations)

$$\begin{aligned} \left| E_{g_0} f_n^0(x) - E_f f_n^0(x) \right| &= \frac{1}{n h_n^d} \left| \sum_{t=1}^{n} \int (\pi^t g_0 - f) \right| \\ &\leq \frac{1}{n h_n^d} \sum_{t=1}^{n} \parallel \pi^t g_0 - f \parallel_\infty h_n^d \end{aligned}$$

then by using D_3 we easily get

$$(2.49) \qquad \left| E_{g_0} f_n^0(x) - E_f f_n^0(x) \right|^2 = O(n^{-2}) \ .$$

On the other hand

$$\widetilde{V}_{g_0} f_n^0 - \widetilde{V}_f f_n^0 = \frac{1}{n^2 h_n^{2d}} \sum_{t=1}^{n} \left[\int_{B_n} \pi^t g_0 \int_{B_n^c} \pi^t g_0 - \int_{B_n} f \int_{B_n^c} f \right]$$

hence

$$\left| \widetilde{V}_{g_0} f_n^0 - \widetilde{V}_f f_n^0 \right| \leq \frac{1}{n^2 h_n^{2d}} \sum_{t=1}^{n} \left[\left| \int_{B_n} \pi^t g_0 - f \right| \left| 1 + \int_{B_n} (\pi^t g_0 + f) \right| \right]$$

$$\leq \frac{3}{n^2 h_n^{2d}} \sum_{t=1}^{n} \| \pi^t g_0 - f \|_\infty$$

thus

$$(2.50) \qquad \left| \widetilde{V}_{g_0} f_n^0 - \widetilde{V}_f f_n^0 \right| = O\left(\frac{1}{n^2 h_n^{2d}} \right) \ .$$

It remains to find a bound for

$$C_n^{(g_0)}(x) = \frac{2}{n^2 h_n^{2d}} \sum_{1 \leq t < t' \leq n} \left| P(X_t \in B_n, X_{t'} \in B_n) - P(X_t \in B_n)P(X_{t'} \in B_n) \right| \ .$$

To this aim we note that if $0 < t < t'$

$$\left| \int_{B_n \cap r^{t'-t}(B_n)} \pi^t g_0 - \int_{B_n \cap r^{t'-t}(B_n)} f \right| \leq \| \pi^t g_0 - f \|_\infty \, h_n^d$$

and that

$$\left| \int_{B_n} \pi^t g_0 \int_{B_n} \pi^{t'} g_0 - \left(\int_{B_n} f \right)^2 \right| \leq 2 \, \| \pi^t g_0 - f \|_\infty \, h_n^d \ ,$$

therefore

$$(2.51) \qquad \begin{aligned} \left| C_n^{(g_0)}(x) - C_n(x) \right| &\leq \frac{6}{n^2 h_n^{2d}} \sum_{t=1}^{n-1} (n-t) \, \| \pi^t g_0 - f \|_\infty \, h_n^d \\ &\leq \frac{6}{n h_n^d} \sum_{t=1}^{+\infty} \| \pi^t g_0 - f \|_\infty \ , \end{aligned}$$

finally (2.46), (2.48), (2.49), (2.50) and (2.51) imply (2.45). ∎

2.5.2 Singular distribution

Consider a stationary Process with marginal distributions concentrated on S, a closed set in $\mathbb{R}^d$ of Lebesgue measure zero. Then we will see that f_n explodes in a neighbourhood of S and vanishes elsewhere. Making use of these properties it is thus possible to construct an estimate for S.

More precisely we consider a GSM strictly stationary $\mathbb{R}^d$-valued process $(Z_t, t \in \mathbb{Z})$ with components $(Y_t, X_t, \ldots, X_{t-k})$ where $k = d - 2 \geq 0$ and such that

(2.52) $$Y_t = r(X_t, \ldots, X_{t-k}) \quad , t \in \mathbb{Z} .$$

We consider the density estimate written under the form

$$f_n(x, y) = \frac{1}{(n-k)h_n^{k+2}} \sum_{t=1}^{n-k} K\left(\frac{x - X_{(t)}}{h_n}, \frac{y - Y_{t+k}}{h_n}\right) ,$$

where $n > k$, $x \in \mathbb{R}^{k+1}$, $y \in \mathbb{R}$, $X_{(t)} = (X_t, \ldots, X_{t-k})$.
For convenience we take $K = 1_{[-\frac{1}{2}, +\frac{1}{2}]^{k+2}}$ and, as usual, we suppose that (h_n) tends to zero.

The assumptions we shall use are the following

S_1) There exists $c_r > 0$ and a neighbourhood V of x such

$$|r(x") - r(x')| \leq c_r \| x" - x' \| \quad ; x', x" \in V .$$

S_2) $X_{(t)}$ has a density f_X continuous and strictly positive at x.

Then the zero-infinite behaviour of f_n is given by the next theorem.

THEOREM 2.5

1) *If* S_1 *holds then for* $n \geq n_0$ *where* n_0 *depends only on* (h_n) *we have*

(2.53) $$\sup_{|y - r(x)| > \max(1, c_r) h_n} f_n(x, y) = 0.$$

2) *If* S_1 *and* S_2 *hold and if* $\dfrac{n h_n^{k+1}}{(\mathrm{Log} n)^2} \to +\infty$ *then for all* $\gamma \in \left]0, \dfrac{1}{2}\right[$ *we have*

(2.54) $$\varliminf h_n \inf_{|y - r(x)| < \gamma h_n} f_n(x, y) > 0 \quad a.s. \quad .$$

It is easy to see that similar results may be proved for other kernel estimates (see [BO1]).

Proof

1) First we may and do suppose that $c_r \geq 1$ and that we use the sup norms in $\mathbb{R}^{k+1}$ and $\mathbb{R}^{k+2}$. Now if n is large enough we may use S_1 to write that

$$\| X(t) - x \| < \frac{h_n}{2} \text{ implies } |r(X_{(t)}) - r(x)| \leq c_r \frac{h_n}{2} \ .$$

Thus, if $|y - r(x)| > c_r h_n$ we have

$$
\begin{aligned}
c_r h_n |y - r(x)| &\leq & |y - r(X_{(t)})| &+& |r(X_{(t)}) - r(x)| \\
&\leq & |y - r(X_{(t)})| &+& c_r \frac{h_n}{2}
\end{aligned}
$$

which entails $|y - r(X_{(t)})| > c_r \dfrac{h_n}{2}$.

Consequently we have

(2.55) $\qquad \Omega = \left\{ \| X_{(t)} - x \| \geq \dfrac{h_n}{2} \right\} \bigcup \left\{ |y - r(X(t))| > c_r \dfrac{h_n}{2} \right\}$

Now put

$$u_{nt}(x,y) = \left(\frac{x - X_{(t)}}{h_n}, \frac{y - r(X_{(t)})}{h_n} \right) \ , \ 1 \leq t \leq n - k \ ,$$

by using (2.55) we get

$$\| u_{nt}(x,y) \| > \frac{1}{2} \min(1, c_r) = \frac{1}{2} \ , 1 \leq t \leq n - k$$

thus we obtain $f_n(x,y) = 0$ hence (2.53). ∎

2) From $|y - r(x)| < \gamma h_n$ with $0 < \gamma < \frac{1}{2}$ and $\| x - X_{(t)} \| \leq a h_n$ we deduce that for n large enough, we have

$$|y - r(X_{(t)})| \leq (\gamma + a c_r) h_n \leq \frac{h_n}{2} \ .$$

Now, if $a = \min \left(\dfrac{1 - 2\gamma}{6 c_r}, \dfrac{1}{2} \right)$, we obtain $\| u_{nt}(x,y) \| \leq \dfrac{1}{2}$ which implies $K(u_{nt}(x,y)) = 1$ for $t = 1, \ldots, n - k$.

Therefore for all y such that $|y - r(x)| < \gamma h_n$, we have

(2.56) $\qquad (2a)^{-k-1} h_n f_n(x,y) \geq f_n^1(x)$

where

$$f_n^{1\prime}(x) = \frac{1}{(n-k)h_n^{k+1}} \sum_{t=1}^{n-k} (2a)^{-k-1} \mathbf{1}_{\|x - X_{(t)}\| \leq a h_n} \ .$$

Now it suffices to note that f_n^1 is a kernel density estimate, then, applying Lemma 2.1 we obtain

$$f_n^1(x) \longrightarrow f_X(x) \quad \text{a.s.}$$

and by (2.56)

$$\underline{\lim} \inf_{|y-r(x)|<\gamma h_n} (2a)^{-k-1} h_n f_n(x,y) \geq f_X(x) \quad \text{a.s.}$$

which is the claimed result. ■

An an application we now consider an estimate r_n of r defined by

$$(2.57) \qquad f_n(x, r_n(x)) = \max_{y \in \mathbb{R}} f_n(x, y)$$

where $r_n(x)$ is chosen in such a way that it should be measurable.

The consistency of r_n is given by the following

COROLLARY 2.3 *If* S_1 *and* S_2 *are satisfied and if* $\dfrac{nh_n^{k+1}}{(\text{Log} n)^2} \to +\infty$ *then for each* $\delta \in]0, 1[$

$$(2.58) \qquad h_n^{-1+\delta}(r_n(x) - r(x)) \longrightarrow 0 \quad a.s. \; .$$

Proof

Let ε be a positive number. If $h_n^{-1+\delta}|r_n(x) - r(x)| > \varepsilon$ then we have $|r_n(x) - r(x)| > \max(1, c_r)h_n$, $n \geq n_\varepsilon$ using (2.53) we obtain for $n \geq \max(n_0, n_\varepsilon)$

$$f_n(x, r_n(x)) = 0$$

and from (2.57)

$$f_n(x, y) = 0 \; , \; y \in \mathbb{R} \; , \; n \geq \max(n_0, n_\varepsilon) \; .$$

Now (2.54) implies that this event has a probability zero, hence the result. ■

We refer to [BO1] for the construction of an estimate for κ which uses again Theorem 2.5.

2.5.3 Processes with errors in variable.

Let $(X_t, t \in \mathbb{Z})$ be a real strictly stationary process. Suppose that each X_t is observed with a random error ε_t, thus the observed process can be written as

$$Y_t = X_t + \varepsilon_t \; , \; t \in \mathbb{Z}$$

where the ε_t's are i.i.d. and the processes (X_t) and (ε_t) are independent.

Here we have a **Deconvolution Problem** : estimate the density f of X_t from the data $Y_1, \ldots, Y_n$, the density f_ε of ε_t being known. That last assumption is somewhat restrictive but it ensures the identifiability of the problem.

The commonly used estimate is given by

$$(2.59) \qquad \hat{f}_n(x) = \frac{1}{nh_n} \sum_{t=1}^{n} \tilde{K}_n\left(\frac{x - Y_t}{h_n}\right) \quad , \; x \in \mathbb{R}$$

where $\tilde{K}_n$ is the deconvolution kernel :

$$(2.60) \qquad \tilde{K}_n(y) = \frac{1}{2\pi} \int_{\mathbb{R}} e^{-iuy} \frac{\phi_K(u)}{\phi_\varepsilon\left(uh_n^{-1}\right)} du \quad , \; y \in \mathbb{R} \, .$$

In (2.60) ϕ_ε denotes the characteristic function of ε_0 and $\tilde{\phi}_K$ the Fourier transform of the classical kernel K.

Under some technical assumptions, asymptotic results may be derived. Clearly the convergence rates are weaker than those in the regular case. Here we only give some indications, we refer to the literature for a complete exposition.

Two important cases may be considered

1) If ϕ_ε is **algebraically decreasing** at infinite it is possible to reach good convergence rates.

   In fact under some regularity and strongly mixing conditions and if $f \in C_{2,d}(b)$ it may be proved that

$$(2.61) \qquad n^{\frac{4}{5+2\beta}} E(\hat{f}_n(x) - f(x))^2 \longrightarrow c > 0$$

   where c is specified and where β is such that

$$\lim_{|u| \to \infty} u^\beta \phi_\varepsilon(u) = a \neq 0 \, .$$

   On the other hand we have

$$(2.62) \qquad \sup_{x \in D} |\hat{f}_n(x) - f(x)| = O\left(\left(\frac{\text{Log}n}{n}\right)^{\frac{2}{5+2\beta}}\right) \quad \text{a.s;}$$

   where D is a compact set.

   Then the loss is small : compare (2.61) with (2.7) and (2.62) with (2.25).

2) If ϕ_ε is **geometrically decreasing** at infinite the convergence rates are poor.

If, for instance, $\phi_\varepsilon(u)$ is of order $\exp(-a|u|^\beta)$ $(a > 0)$ at infinite then under some conditions

$$E(\hat{f}_n(x) - f(x))^2 = O((\text{Log}\,n)^{-4/\beta})$$

and

$$\sup_{x \in D} |\hat{f}_n(x) - f(x)| = O((\text{Log}\,n)^{-2/\beta})$$

where D is compact.

Unfortunately an improvement is not possible since the above rates and the last two ones specifically are optimal (see [FA1]).

For multidimensional versions of the above results and asymptotic normality we refer to the bibliography.

Notes

The kernel density estimator was introduced by ROSENBLATT ([RO2]) in 1956. A great number of people have studied this estimate in the i.i.d. case.

In the strongly mixing case one can mention ROUSSAS, ROSENBLATT, TRAN, TRUONG-STONE, MASRY, BOSQ, ROBINSON, PHAM-TRAN among others.

Chaotic data and singular distributions are studied by BOSQ ([BO1] and [BO2]). Processes with errors have been recently considered by FAN and MASRY.

The choice of bandwidth will be discussed in the next chapter.

Chapter 3

Regression estimation and prediction for discrete time processes

The construction and study of a nonparametric predictor are the main purpose of this chapter. In practice such a predictor is in general more efficient and more flexible than the predictors based on BOX and JENKINS method, and nearly equivalent if the underlying model is truly linear. This surprising fact will be clarified at the end of the chapter.

In Sections 1 and 2 we will study the kernel regression estimator obtaining optimal rates in quadratic mean and uniformly almost surely and deriving the asymptotic distribution. Section 3 will be devoted to prediction for a kth order Markov process. Prediction in the general case will be presented in Section 4. This section also contains some ideas about related topics : interpolation, outliers detection, chaos, regression with error. Finally the implementation of the kernel method and its comparison with the classical linear method are discussed in Section 5.

3.1 Regression estimation

Let $Z_t = (X_t, Y_t), t \in \mathbb{Z}$ be a $\mathbb{R}^d \times \mathbb{R}^{d'}$-valued *strictly stationary* process and let m be a Borelian function of $\mathbb{R}^{d'}$ into $\mathbb{R}$ such that $E(|m(Y_0)|) < +\infty$.

We suppose that Z_0 admits a density $f_Z(x, y)$ and that $f_Z(x, \cdot)$ and $m(\cdot)$

are in $\mathcal{L}^1(\lambda^{d'})$ for *each* x in $\mathbb{R}^d$. Then, we may define the functional parameters

(3.1) $$f(x) = \int f_Z(x,y)dy \quad , \quad x \in \mathbb{R}^d.$$

(3.2) $$\varphi(x) = \int m(y)f_Z(x,y)dy \quad , \quad x \in \mathbb{R}^d.$$

and

$$
\begin{aligned}
r(x) &= \varphi(x)/f(x) && \text{if} \quad f(x) > 0 \\
&= Em(Y_0) && \text{if} \quad f(x) = 0.
\end{aligned}
$$
(3.3)

Clearly, $r(\cdot)$ is a version of $E(m(Y_0)|X_0 = \cdot)$. We will say that $r(\cdot)$ is a **regression parameter**.

Typical example of regression parameters are

$$
\begin{aligned}
r(x) &= P(Y \in B \mid X = x) &&, \quad (B \in \mathcal{B}_{\mathbb{R}^d}) \\
r(x) &= E(Y^k \mid X = x) &&, \quad (k \geq 1, d' = 1) \\
r(x) &= V(Y \mid X = x) &&, \quad (d' = 1)
\end{aligned}
$$

The problem is to construct an estimator of r based on the data $Z_t,\ 1 \leq t \leq n$. As for density the method uses a convolution kernel which regularizes the empirical measures.

Consider the empirical measure

$$\nu_n = \frac{1}{n}\sum_{t=1}^n \delta_{(X_t, m(Y_t))}$$

and its marginal distribution

$$\mu_n = \frac{1}{n}\sum_{t=1}^n \delta_{(X_t)}.$$

A regularization of μ_n and ν_n by convolution leads to natural estimators of f and φ :

(3.4) $$f_n(x) = \frac{1}{nh_n^d}\sum_{t=1}^n K\left(\frac{x - X_t}{h_n}\right) \quad , \quad x \in \mathbb{R}^d$$

and

(3.5) $$\varphi_n(x) = \frac{1}{nh_n^d}\sum_{t=1}^n m(Y_t)K\left(\frac{x - X_t}{h_n}\right) \quad , \quad x \in \mathbb{R}^d ,$$

where K is a strictly positive *kernel* (see Chapter 2) and h_n a smoothing parameter satisfying $\lim_{n\to\infty} h_n = 0(+)$.

Practical considerations leading to r_n are discussed in S.3.

Consequently the *kernel estimator* of r is defined as

$$(3.6) \qquad r_n(x) = \varphi_n(x)/f_n(x) , \quad x \in \mathbb{R}^d.$$

Note that, if K is not strictly positive, definition (3.6) must be completed : for $f_n(x) = 0$ one may choose $r_n(x) = \dfrac{1}{n}\sum_{t=1}^{n} m(Y_t)$ which is clearly more natural than the arbitrary $r_n(x) = 0$ used by many authors.

Note also that an interesting form of r_n should be the weighted mean

$$(3.7) \qquad r_n(x) = \sum_{t=1}^{n} p_{nt}(x)m(Y_t)$$

where $p_{nt}(x) = K\left(\dfrac{x - X_t}{h_n}\right) / \sum_{s=1}^{n} K\left(\dfrac{x - X_s}{h_n}\right)$, $1 \leq t \leq n$ if $f_n(x) \neq 0$ and $p_{nt}(x) = \dfrac{1}{n}, 1 \leq t \leq n$ if $f_n(x) = 0$.

3.2 Asymptotic behaviour of the regression estimator

3.2.1 Quadratic error

In order to obtain the exact asymptotic quadratic error for r_n, we need the following assumptions and notations :

1. The density $f_{(X_s,X_t)}$ exists for $s \neq t$, belongs to $C_{2,d}(b)$ and satisfies

$$\sup_{s\neq t} \| f_{(X_s,X_t)} - f \otimes f \|_p < +\infty$$

   for some $p \in]2, +\infty]$.

2. f and φ belong to $C_{2,d}(b)$, f_Z belongs to $C_{2,d+d'}(b)$ for some b.

3. There exists $a > 0$ such that

$$E(\exp a|m(Y_t)|) < +\infty .$$

4. The strong mixing coefficient $\alpha(\cdot)$ of (Z_t) satisfies

$$\alpha(k) \leq \gamma k^{-\beta} \ , \ \ k \geq 1$$

where $\gamma > 0$ and $\beta > 0$.

Now, if $f(x) > 0$, we set

$$C(x, c, K, f, r) \ = \ \frac{c^4}{4} \left(\sum_{1 \leq i, j \leq d} \frac{\partial^2 r}{\partial x_i \partial x_j} + \frac{\partial \mathrm{Log} f}{\partial x_i} \frac{\partial r}{\partial x_j} \int u_i u_j K(u) du \right)^2$$
$$+ \ c^{-d} \frac{v(x)}{f(x)} \int K^2 \ ,$$

where $v(x) = \int m(y)^2 \dfrac{g(x,y)}{f(x)} dy - r(x)^2$ is a version of $V(m(Y_0)|X_0 = x)$.

Then we have

THEOREM 3.1 *If $1 \to 4$ hold, if $f(x) > 0$ and if $\beta > \max(\dfrac{2(p-1)}{p-2}, d+2)$ then the choice $h_n = c_n n^{-1/(d+4)}$ where $c_n \to c > 0$ leads to*

$$(3.8) \qquad n^{4/(d+4)} E(r_n(x) - r(x))^2 \to C(x, c, K, f, r).$$

Proof

We omit x and write

$$(3.9) \qquad r_n - r = (r_n - r)\frac{f - f_n}{f} + \frac{r}{f}(f - f_n) + \frac{\varphi_n - \varphi}{f},$$

thus

$$E(r_n - r)^2 = A_n + B_n + C_n$$

where

$$A_n = \frac{r^2}{f^2} E(f - f_n)^2 + \frac{1}{f^2} E(\varphi_n - \varphi)^2 + \frac{2r}{f^2} E[(f - f_n)(\varphi_n - \varphi)]$$

$$B_n = \frac{1}{f^2} E\left[(r_n^2 - r^2)(f_n - f)^2\right] \ ,$$

and

$$C_n = -\frac{2}{f^2} E[(r_n - r)(\varphi_n - \varphi)(f_n - f)].$$

The quantity $E(f - f_n)^2$ has been already studied in Theorem 2.1 and the other terms in A_n may be studied similarly. After some calculations one obtains

$$n^{4/(d+4)} A_n \to C(x, c, K, f, r) .$$

It remains to prove the asymptotic negligibility of B_n and C_n with respect to $n^{-4/(d+4)}$. We only consider B_n since the treatment of C_n is similar.

Given $\gamma > 0$ and $\varepsilon > 0$, we have

$$B_n \leq (n^\gamma + |r|) E(|r_n - r| \mathbf{I}_{|r_n| \leq n^\gamma})(f_n - f)^2 \left(\mathbf{I}_{|r_n - r| \leq n^{(-1-\varepsilon)\gamma}} \right. \\ \left. + \mathbf{I}_{|r_n - r| > n^{(-1-\varepsilon)\gamma}} \right) .$$

Using successively Schwarz and Hölder inequalities we get for n large enough :

$$B_n \leq 2n^\gamma n^{-(1+\varepsilon)\gamma} E(f_n - f)^2 + 2n^\gamma [E(f_n - f)^4]^{1/2}$$

(3.10)

$$\left[E \left(|r_n - r|^{2v} \mathbf{I}_{|r_n| \leq n^\gamma} \right) \right]^{1/2v} \left[P \left(|r_n - r| > n^{-(1+\varepsilon)\gamma}, |r_n| \leq n^\gamma \right) \right]^{1/2v} ,$$

where $\dfrac{1}{v} + \dfrac{1}{w} = 1$.

The first term in the bound is a $O(n^{-\varepsilon\gamma} n^{-4/(d+4)}) = o(n^{-4/(d+4)})$. In the second term we treat each factor separately. First we clearly have

$$\left(E(f_n - f)^4 \right)^{1/2} \leq \| K \|_\infty h_n^d \left(E(f_n - f)^2 \right)^{1/2}$$

(3.11)

$$= O \left(n^{\frac{d-2}{d+4}} \right) .$$

On the other hand since

$$E \left((r_n - r)^2 \mathbf{I}_{|r_n| \leq n^\gamma} \right) \leq \frac{2n^{2\gamma}}{f^2} E(f_n - f)^2 + \frac{2}{f^2} E(\varphi_n - \varphi)^2$$

$$= O \left(n^{2\gamma - \frac{4}{d+4}} \right)$$

we have, writing $2v = 2(v-1) + 2$,

$$E \left(|r_n - r|^{2v} \mathbf{I}_{|r_n| \leq n^\gamma} \right) = O \left(n^{\gamma \frac{2(v-1)}{2v} + \left(2\gamma - \frac{4}{d+4} \right) \frac{1}{2v}} \right)$$

(3.12)

$$= O \left(n^{\gamma - \frac{2}{v(d+4)}} \right) .$$

In order to evaluate the last factor, we put

$$V_t = h_n^{-d} K \left(\frac{x - X_t}{h_n} \right) m(Y_t) \mathbf{I}_{|m(Y_t)| \leq n^\gamma}$$

and

$$W_t = h_n^{-d} K \left(\frac{x - X_t}{h_n} \right) m(Y_t) \mathbf{I}_{|m(Y_t)| > n^\gamma} \ .$$

First we have

$$E_n =: \left\{ \frac{1}{n} \frac{1}{f} \left| \sum_{t=1}^{n} (W_t - EW_t) \right| > \frac{n^{-(1+\varepsilon)\gamma}}{8} \right\} \Rightarrow \bigcup_{t=1}^{n} \{ |m(Y_t)| > n^\gamma \}.$$

then, assumption 3 and Markov inequality imply

$$P(E_n) \leq n \exp(-an^\gamma) E(\exp a |m(Y_0)|) \ .$$

On the other hand we may apply the exponential type inequality (1.34) to the random variables $V_t - EV_t$, $1 \leq t \leq n$. Then, choosing $q \simeq n^A$, collecting the above bounds and choosing v large enough, γ and ε small enough and A close enough to $\dfrac{4}{d+4}$ we get $B_n = o(n^{4/(d+4)})$ and the proof of Theorem 3.1 is now complete. ∎

3.2.2   Uniform almost sure convergence

Uniform convergence of a regression estimator may be obtained over compact sets, but in general not over the whole space, even if some information about the behaviour of $r(x)$ for $\| x \|$ large is available.

As a simple example let us consider the case where the Z_t's are i.i.d. bivariate Gaussian variables with standard margins and $\mathrm{Cov}(X_t, Y_t) = \rho$. In this case the classical estimator of $r(x) = \rho x$ is $\rho_n x$ where

$$\rho_n = \frac{1}{n} \sum_{t=1}^{n} X_t Y_t$$

and clearly

(3.13) $\sup_{x \in \mathbb{R}} |\rho_n x - \rho x| = +\infty$ a.s..

In fact it is impossible to construct a regression estimator R_n such that

$$\sup_{x \in \mathbb{R}} |R_n(x) - \rho x| \to 0 \ \ a.s.$$

since this property implies that, wheterver $(u_n) \uparrow \infty$,

$$u_n \left(\frac{R_n(u_n)}{u_n} - \rho \right) \to 0 \ \ a.s.$$

and consequently it should be possible to obtain an estimator of ρ with an arbitrary sharp rate of convergence.

However it is possible to establish uniform convergence over suitable increasing sequences of compact sets. In the following, we will say that a sequence (S_n) of compact sets in $\mathbb{R}^d$ is *regular* (with respect to f) if there exists a sequence (β_n) of real numbers and a $\gamma > 0$ such that for each n

$$\inf_{x \in S_n} f(x) \geq \beta_n > 0 \text{ and } \delta(S_n) \leq n^{\gamma}$$

where $\delta(S_n)$ denotes the diameter of S_n.

We first consider convergence on a fixed compact set. In that case the obtained rate is optimal (up to a logarithm).

THEOREM 3.2 *Let (Z_t) be a GSM strictly stationary process such that f and φ belong to $C_{2,d}(b)$ for some b and such that $E(\exp a|m(Y_0)|^{\tau}) < +\infty$ for some $a > 0$ and some $\tau > 0$.*

Then if K is Lipschitzian, if $\dfrac{nh_n^d}{(\mathrm{Log}\, n)^{2+\frac{1}{\tau}}} \to +\infty$ and if S is a compact set such that $\inf_{x \in S} f(x) > 0$, we have

(3.14) $$\sup_{x \in S} |r_n(x) - r(x)| \to 0 \quad a.s..$$

Furthermore if $h_n \simeq \left(\dfrac{(\mathrm{Log}\, n)^{2-\frac{1}{\tau}}}{n} \right)^{1/(d+4)}$, then for each integer k

(3.15) $$\frac{n^{2/(d+4)}}{\mathrm{Log}_k n (\mathrm{Log}\, n)^{\frac{1}{\tau} + (1-\frac{1}{\tau})\frac{2}{d+4}}} \sup_{x \in S} |r_n(x) - r(x)| \to 0 \quad a.s. \ .$$

Proof (Sketch)

We omit x and write

$$\sup_S |r_n - r| \leq \frac{\sup_S |r_n|}{\inf_S f} \sup_S |f_n - f| + \frac{1}{\inf_S f} \sup_S |\varphi_n - \varphi| \ .$$

First, by using (3.7) we get for each $A > 0$

$$P(\sup_S |r_n| > A) \leq P(\sup_{1 \leq t \leq n} |m(Y_t)| > A) \ .$$

Now, Markov inequality entails

$$P(\sup_S |r_n| > A) \leq n \exp(-aA^{\tau}) E(\exp a|m(Y_0)|^{\tau}),$$

then, choosing $A = c^\tau (\mathrm{Log} n)^{1/\tau}$ where $c > \dfrac{2}{a}$ and using Borel Cantelli lemma we obtain

$$P(\limsup_n \{\sup_S |r_n| > c^\tau (\mathrm{Log} n)^{1/\tau}\}) = 0.$$

On the other hand, using inequality (1.25) and a covering of S it is easy to prove that $\sup_S |\varphi_n - \varphi| \to 0$ a.s. and $(\mathrm{Log} n)^{1/\tau} \sup_S |f_n - f| \to 0$ a.s. (see the proof of Theorem 2.2), hence (3.14).

The proof of (3.15) is similar but uses the more precise inequality (1.26) instead of (1.25). Details are omitted. ∎

The following theorem considers a varying compact set :

THEOREM 3.3 *If assumptions of Theorem 3.2 hold, if (δ_n) is a sequence of real numbers such that there exists a regular sequence of compact sets satisfying*

$$\frac{\delta_n (\mathrm{Log} n)^{(1+\frac{1}{2\tau})-(1-\frac{1}{2\tau})\frac{2}{d+4}}}{\beta_n n^{2/(d+4)}} \longrightarrow 0$$

then the choice $h_n \simeq \left(\dfrac{(\mathrm{Log} n)^{2-(1/\tau)}}{n} \right)^{1/(d+4)}$ entails

(3.16) $\delta_n \cdot \sup\limits_{x \in S_n} |r_n(x) - r(x)| \to 0 \quad a.s..$

The proof is similar to that of Theorem 3.2 and therefore omitted.

Example 3.1

Suppose that $m = \mathbf{1}_B$ where $B \in \mathcal{B}_{\mathbb{R}^d}$ and that $f(x) \simeq \| x \|^{-p}$ for $\| x \|$ large enough $(p > d)$, then if $\delta < \dfrac{2}{d+4} - \gamma p$ we have

$$n^\delta \sup_{\|x\| \leq n^\gamma} \left| \sum_{t=1}^n p_{nt}(x) \mathbf{1}_B(Y_t) - P(Y_0 \in B | X_0 = x) \right| \to 0 \text{ a.s.}$$

where $p_{nt}(x)$ is defined with (3.7).

Example 3.2

Suppose that (Z_t) is a bivariate Gaussian process with $X_t \sim \mathcal{N}(0, \sigma^2)$ $(\sigma > 0)$. Then, if m is the identity, we have for each $\varepsilon \in]0, \frac{2}{5}[$

$$\frac{n^{\frac{2}{5}-\varepsilon}}{\mathrm{Log} n} \sup_{|x| \leq \sigma \sqrt{2\varepsilon \mathrm{Log} n}} |r_n(x) - r(x)| \to 0 \text{ a.s..}$$

3.2.3 Limit in distribution

We now state some conditions which ensure weak convergence of $\gamma_n(r_n(x) - r(x))$ for a suitable choice of (γ_n).

For convenience we suppose that

$$Z_t = (X_t, Y_t) = ((\xi_t, \dots, \xi_{t+k-1}), \, \xi_{t+k-1+H}), \, t \in \mathbb{Z}$$

where $(\xi_t, t \in \mathbb{Z})$ is a real strictly stationary process, and that m is the identity.

Now the assumptions are

1. $E|X_0|^{4+\delta} < +\infty$ for some positive δ.

2. (X_t) is α-mixing with $\alpha(k) = O(k^{-\beta})$ where $\beta > \dfrac{\delta^2 + 4}{2\delta}$.

3. f and φ are in $C_{2,1}(b)$ for some b.

4. $f(\cdot)E\left(Y_0^2|X_0 = \cdot\right)$ is continuous at x.

5. $\sup_{t \in \mathbb{Z}} \| E\left(Y_0^i Y_t^j | X_t = \cdot, \, X_0 = \cdot\right) \|_\infty < +\infty$

   where $i \geq 0, j \geq 0, i + j = 2$.

6. $\sup_{t \geq k} \| f_{(X_0, X_t)} \|_\infty < +\infty$.

7. $\| E\left(|Y_0|^{4+\delta}|X_0 = \cdot\right) \|_\infty < +\infty$

then we have

THEOREM 3.4 *If conditions* $1 \longrightarrow 7$ *hold, if* $f(x)v(x) > 0$ *and if* $nh_n^{d+2} \longrightarrow 0$ *then*

$$(3.17) \qquad \left(\int K^2\right)^{-1} \sqrt{nh_n^d} \, \sqrt{\frac{f_n(x)}{v_n(x)}} \, (r_n(x) - r(x)) \overset{w}{\longrightarrow} N \sim \mathcal{N}(0, 1)$$

where v_n *is the kernel estimator of* v.

The proof is rather intricate but similar to the proof of Theorem 2.3 and therefore omitted.

3.3 Prediction for a stationary Markov process of order k

Let $(\xi_t, t \in \mathbb{Z})$ be a $\mathbb{R}^{d_0}$-valued strictly stationary process. Suppose that (ξ_t) is a Markov process of order k, namely

$$(3.18) \qquad \mathcal{L}(\xi_t \mid \xi_{t-s}, s \geq 1) = \mathcal{L}(\xi_t \mid \xi_{t-1}, \ldots, \xi_{t-k}), \quad \text{a.s.}$$

or equivalently

$$E\left(F(\xi_t) \mid \xi_{t-s}, s \geq 1\right) = E\left(F(\xi_t) \mid \xi_{t-1}, \ldots, \xi_{t-k}\right) \quad \text{a.s.}$$

for each Borelian real function F such that $E(|F(\xi_0)|) < +\infty$.

Given the data $\xi_1, \ldots, \xi_N$ we want to predict the non-observed square integrable real random variable

$$\zeta_{N+H} = m(\xi_{N+H})$$

where $1 \leq H \leq N-k$ and where m is measurable and bounded on compact sets.

For that purpose let us construct the *associated process*

$$Z_t = (X_t, Y_t) = ((\xi_t, \ldots, \xi_{t+k-1}), m(\xi_{t+k-1+H})), \ t \in \mathbb{Z},$$

and consider the kernel regression estimator r_n based on the data $(Z_t, 1 \leq t \leq n)$ where $n = N - k + 1 - H$. In the present case $d = kd_0$ and $d' = d_0$.

From r_n we construct the predictor

$$(3.19) \qquad \widehat{\zeta}_{N+H} = r_n(X_{n+H}) \,,$$

and we set

$$r(x) = E(m(\xi_{k-1+H}) \mid (\xi_0, \ldots, \xi_{k-1}) = x) \,, \ x \in \mathbb{R}^d.$$

3.3.1 Quadratic prediction error

We study here the asymptotic quadratic prediction error first for a bounded process and then in the general case. In the bounded case, we have the following

THEOREM 3.5 *If (ξ_t) is a bounded, GSM, strictly stationary k^{th} order Markovian process and if the associated process (Z_t) admits functional parameters f and φ continuously differentiable on $S = \{f > 0\}$ then, for a Lipschitzian*

K *and* $h_n \simeq \left(\dfrac{(\mathrm{Log} n)^{2-\varepsilon}}{n} \right)^{1/(d+2)}$ *where*

$0 < \varepsilon < 2$ *we have*

$$(3.20) \qquad E(r_n(X_{n+H}) - r(X_{n+H}))^2 = O\left(\frac{(\mathrm{Log} n)^{2\varepsilon+(2-\varepsilon)\frac{2}{d+2}}}{n^{2/(d+2)}} \right).$$

Proof (sketch)

First we clearly have

$$E(r_n(X_{n+H}) - r(X_{n+H}))^2 \leq E\left[\sup_{x \in S}(r_n(x) - r(x))^2 \right].$$

Now, an integration by parts gives

$$E(\sup_{x \in S}(r_n(x) - r(x))^2) = 2 \int_0^{+\infty} v P(\sup_{x \in S} |r_n(x) - r(x)| > v) dv$$

Finally, in order to bound above the integrand, it suffices to make a covering of S and then use the exponential inequality (1.26) (see the proofs of Theorems 2.2 and 3.2). ∎

Note that if f and φ are twice continuously differentiable on S the rate is not improved because the bias remains the same on the edge of S.

We deal with the general case in the following

THEOREM 3.6 *If* (ξ_t) *is a GSM, strictly stationary, k^{th} order Markovian process and if the associated process* (Z_t) *satisfies : f and $\varphi \in C_{2,d}(b)$ for some b and* $E(\exp a|Y_t|^\tau) < \infty$ *for some $a > 0$ and some $\tau > 0$; then for a Lipschitzian K,* $h_n \simeq \left(\dfrac{(\mathrm{Log} n)^{2-\frac{1}{\tau}}}{n} \right)^{1/(d+4)}$ *and each regular sequence* (S_n) *of compact sets*

$$(3.21) \qquad E\left((r_n(X_{n+H}) - r(X_{n+H}))^2 \mathbf{1}_{X_{n+H} \in S_n} \right) =$$
$$= O\left(\frac{(\mathrm{Log} n)^{\frac{2}{\tau}+(2-\frac{1}{\tau})\frac{4}{d+4}}}{\beta_n^2 n^{4/(d+4)}} \right)$$

and

$$(3.22) \qquad E\left((r_n(X_{n+H}) - r(X_{n+H}))^2 \mathbf{1}_{X_{n+H} \notin S_n} \right) =$$
$$= O\left((\mathrm{Log} n)^{2/\tau} P(\| X_0 \| > \delta(S_n)^{1/2}) \right).$$

Proof

The proof of (3.21) is similar to that of (3.20) and is therefore omitted.

Concerning (3.22) first, note that, since conditional expectation is a contraction in $L^4(\Omega, \mathcal{A}, P)$ (see [RA]) we have

$$(3.23) \qquad\qquad Er^4(X_{n+H}) \leq Em^4(\xi_{N+H}) = O(1).$$

On the other hand (3.7) yields

$$r_n(X_{n+H}) \leq \sup_{1 \leq t \leq n} |m(\xi_{t+k-1+H})|$$

and it is easy to prove that the condition $E(\exp a|m(\xi_0)|^\tau) < +\infty$ implies

$$(3.24) \qquad\qquad E\left(\sup_{1 \leq t \leq n} |m(\xi_{t+k-1+H})|^p\right) = O((\mathrm{Log} n)^{p/\tau}),$$

thus
$$(3.25) \qquad\qquad Er_n^4(X_{n+H}) = O((\mathrm{Log} n)^{4/\tau}).$$

Now let us write

$$
\begin{aligned}
\Delta_n &= E\left((r_n(X_{n+H}) - r(X_{n+H}))^2 \mathbf{1}_{X_{n+H} \notin S_n}\right) \\
&\leq 2E\left(r_n^2(X_{n+H})\mathbf{1}_{X_{n+H} \notin S_n}\right) + 2E\left(r^2(X_{n+H})\mathbf{1}_{X_{n+H} \notin S_n}\right)
\end{aligned}
$$

then by using Schwarz inequality, (3.23) and (3.25) we get

$$\Delta_n = O\left((\mathrm{Log} n)^{2/\tau}(P(X_{n+H} \notin S_n)^{1/2})\right)$$

hence (3.22). ∎

Note that while the rate in (3.20) is nearly optimal, there is a loss of rate in the general case. As indicated above, the reason for it is the unpredictable behaviour of $r(x)$ for large values of $\| x \|$.

Example 3.3

Take a one dimensional Markov process (ξ_t) with $f(x) \simeq c \exp(-c'|x|^\tau)$ $(\tau \geq 1, c > 0, c' > 0)$, then, using (3.21) and (3.22) it is easy to check that

$$(3.26) \qquad E(r_n(X_{n+H}) - r(X_{n+H}))^2 = O\left(\frac{(\mathrm{Log} n)^{\frac{2}{\tau} + \frac{4}{5}(2 - \frac{1}{\tau})}}{n^{4/25}}\right).$$

3.3.2 Almost sure convergence of the predictor

The empirical error $|r_n(X_{n+H}) - r(X_{n+H})|$ gives a good idea of the predictor's accuracy. We now study its asymptotic behaviour. As above we separate the bounded case and the general case.

COROLLARY 3.1 *If conditions in Theorem 3.5 hold, then, for each $\varepsilon > 0$,*

$$(3.27) \qquad \frac{n^{1/(d+2)}}{(\mathrm{Log}\,n)^{\varepsilon + (1-\varepsilon)\frac{1}{d+2}}} |r_n(X_{n+H}) - r(X_{n+H})| \longrightarrow 0 \quad a.s..$$

Proof

Since

$$|r_n(X_{n+H}) - r(X_{n+H})| \le \sup_{x \in S} |r_n(x) - r(x)|$$

the result follows from Theorem 3.2 applied to the associated process (Z_t). ∎

COROLLARY 3.2 *If conditions in Theorem 3.6 hold and if*

$$\frac{\delta_n (\mathrm{Log}\,n)^{\left(1+\frac{1}{2\tau}\right) - \left(1-\frac{1}{2\tau}\right)\frac{2}{d+4}}}{\beta_n n^{2/(d+4)}} \longrightarrow 0$$

then

$$(3.28) \qquad \delta_n (r_n(X_{n+H}) - r(X_{n+H})) \mathbf{I}_{X_{n+H} \in S_n} \longrightarrow 0 \quad a.s. \ .$$

Furthermore, if

$$\delta_n' (\mathrm{Log}\,n)^{1/\tau} (P(\| X_0 \| > \delta(S_n))^{1/4} \longrightarrow 0$$

then

$$(3.29) \qquad \delta_n' (r_n(X_{n+H}) - r(X_{n+H})) \mathbf{I}_{X_{n+H} \notin S_n} \xrightarrow{P} 0.$$

Proof

For (3.28) it suffices to write

$$\delta_n |r_n(X_{n+H}) - r(X_{n+H})| \mathbf{I}_{X_{n+H} \in S_n} \le \sup_{x \in S_n} |r_n(x) - r(x)|$$

and then to apply Theorem 3.3.

A simple application of Tchebychev inequality and (3.22) entails (3.29). ∎

In example 3.3 we obtain, up to a logarithm, the rate $n^{-2/25}$.

3.3.3   Limit in distribution

In order to derive the asymptotic distribution of $r_n(X_{n+H})$ we need an independence asymptotic condition stronger than α-mixing : a process $(\xi_t, t \in \mathbb{Z})$ is said to be $\varphi_{rev} - mixing$ if the reversed process $(\xi_{-t}, t \in \mathbb{Z})$ is φ-mixing. For such a process we have the following

LEMMA 3.1 *Let* (ξ_t) *be a* φ_{rev}*-mixing process and let* η *be a* $\sigma(\xi_t, t \leq k)$-*measurable bounded complex random variable, then for each positive integer* p

$$(3.30) \qquad |E(\eta|\xi_{k+p}) - E\eta| \leq 4 \parallel \eta \parallel_\infty \varphi_{rev}(p) \quad a.s.$$

where $\varphi_{rev}(.)$ *is the* φ*-mixing coefficient of* (ξ_{-t}).

Proof See [RS-IO]

THEOREM 3.7 *If* (ξ_t) *is* φ_{rev}*-mixing and if conditions of Theorem 3.4 hold, then*

$$(3.31) \qquad (\int K^2)^{-1} \left(\frac{nh_n^d f_n(X_{n+H})}{v_n(X_{n+H})} \right)^{1/2} [r_n(X_{n+H}) - r(X_{n+H})] \xrightarrow{w} N$$

where $N \sim \mathcal{N}(0, 1)$.

Proof

Let us first consider the kernel estimator

$$r_{n'}(x) = \frac{\displaystyle\sum_{t=1}^{n'} m(Y_t) K \left(\frac{x - X_t}{h_{n'}} \right)}{\displaystyle\sum_{t=1}^{n'} K \left(\frac{x - X_t}{h_{n'}} \right)}, x \in \mathbb{R}^d$$

where $n' = n - [\text{LogLog}n]$ and where $h_{n'} = h_n$.
Let us similarly define $f_{n'}$ and $v_{n'}$, and set

$$Z_{n'}(x) = (\int K^2)^{-1} \left(\frac{n' h_{n'}^d f_{n'}(x)}{v_{n'}(x)} \right)^{1/2} (r_{n'}(x) - r(x)), x \in \mathbb{R}^d.$$

Then it is easy to check that (3.31) is valid if and only if

$$(3.32) \qquad\qquad Z_{n'}(X_{n+H}) \xrightarrow{w} N.$$

Now, in order to establish (3.32), we first apply (3.30) to $e^{iuZ_{n'}(x)}$, obtaining

$$(3.33) \qquad \left| E \left(e^{iuZ_{n'}(x)} | X_{n+H} \right) - E \left(e^{iuZ_{n'}(x)} \right) \right| \leq 4\varphi_{rev}(n + H - n'),$$

and since $E\left(e^{iuZ_{n'}(x)}\right) \longrightarrow e^{-\frac{u^2}{2}}$, we also have

$$E\left(e^{iuZ_{n'}(x)}|X_{n+H}\right) \longrightarrow e^{-\frac{u^2}{2}} \quad \text{a.s..}$$

Now

$$E\left(e^{iuZ_{n'}(X_{n+H})}\right) = \int E\left(e^{iuZ_{n'}(x)}|X_{n+H}=x\right)f(x)dx$$

and the dominated convergence theorem implies

$$E\left(e^{iuZ_{n'}(X_{n+H})}\right) \longrightarrow e^{-\frac{u^2}{2}}, \quad u \in \mathbb{R}$$

which proves (3.32). Theorem 3.7 is therefore established. ∎

Note that, by using the precise form of (3.31), one may construct confidence intervals for $r(X_{n+H})$.

3.4 Prediction for general processes

The assumptions used in the above section allowed us to obtain good rates. However these assumptions are rather restrictive for applications. In the current section we consider some more realistic conditions concerning the observed process. We will successively study the general stationary case, the nonstationary case and some related topics (interpolation, chaos, regression with error).

3.4.1 Prediction for general stationary processes

Most of the stationary processes encountered in practice are not Markovian even if they can be approached by a kth order Markov process for a suitable k. In some cases the process is Markovian but k is unknown. Some methods for choosing k are available in literature, particularly in the linear case : see [BR-DA] and [G-M]. Finally, in practice, k appears as a "truncation parameter" which may depend on the number of observations.

In order to take that fact into account we are induced to consider associated processes of the form

$$Z_{t,N} = (X_{t,N}, Y_{t,N}) = ((\xi_t, \ldots, \xi_{t+k_N-1}), m(\xi_{t+k_N-1+H})), \ t \in \mathbb{Z}, \ N \geq 1,$$

where $\lim_{N\to\infty} k_N = \infty$ and $\lim_{N\to\infty} N - k_N = \infty$. Here the observed process (ξ_t) is $\mathbb{R}^{d_0}$-valued and strictly stationary.

The predictor of $m(\xi_{N+H})$ is defined as

$$(3.34) \qquad r_n^*(X_{N+H,N}) = \frac{\sum_{t=1}^{n} Y_{t,N} K\left(\dfrac{X_{N+H,N} - X_{t,N}}{h_n}\right)}{\sum_{t=1}^{n} K\left(\dfrac{X_{N+H,N} - X_{t,N}}{h_n}\right)}$$

where $n = N - k_N + 1 - H$ and $K = K_0^{\otimes k_N}$ where K_0 is a d_0-dimensional kernel.

Now some martingale considerations imply that $E(m(\xi_{N+H}) \mid \xi_N, \cdots, \xi_{N-k_N+1})$ is close to $E(m(\xi_{N+H})|\xi_s, s \leq N)$ for large N. Then under regularity conditions similar to those of Section 3.3 and using the same methods it may be proved that

$$(3.35) \qquad r_n^*(X_{N+H,N}) - E(m(\xi_{N+H})|\xi_s, s \leq N) \xrightarrow[\text{q.m.}]{\text{a.s.}} 0$$

provided $k_N = O((\text{Log}N)^\delta)$ for some $\delta > 0$.
It is clearly hopeless to reach a sharp rate in the general case. In fact, it can be proved that a $(\text{Log}n)^{-\delta'}$ rate is possible. For precise results and details we refer to [RH].

3.4.2   Prediction for nonstationary processes

We now consider a simple form of nonstationarity supposing that an observed process (η_t) admits the decomposition

$$(3.36) \qquad \eta_t = \xi_t + s_t \ , \ t \in \mathbb{Z}$$

where (ξ_t) is a non-observed strictly stationary process and (s_t) an unknown deterministic sequence. For the estimation of (s_t) we refer to Section 4.

Now, if an estimator $\hat{s}$ of s is available, one may consider the artificial data

$$\hat{\xi}_t = \eta_t - \hat{s}_t \ \ ; \ t = 1, \ldots, n$$

and use it for prediction. However that method suffers from a drawback : $\hat{\xi}_t$ is perturbed and cannot be considered as a good approximation of ξ_t (see for example [G-M]).

Here we only make regularity assumptions on s and do not try to estimate it. In fact we want to show that the nonparametric predictor considered in Section 3.3 exhibits some kind of robustness with respect to the nonstationarity produced by s.

In order to simplify the exposition we assume that (η_t) is a real valued Markov process and that we want to predict η_{n+1} given $\eta_1, \ldots, \eta_n$.

In the following g denotes the density of (ξ_0, ξ_1), f the density of ξ_0, r the regression of ξ_1 on ξ_0 and $\varphi = rf$.

Concerning s we introduce the condition

C - s is bounded and there exist real functions $\overline{f}$ and $\overline{\varphi}$, and a $\delta \geq 0$ such that

$$n^\delta \sup_{x \in \mathbb{R}} \left| \frac{1}{n} \sum_{t=1}^n f(x - s_t) - \overline{f}(x) \right| \xrightarrow[n \to \infty]{} 0$$

and

$$n^\delta \sup_{x \in \mathbb{R}} \left| \frac{1}{n} \sum_{t=1}^n \int yg(x - s_t, y - s_t)dy - \overline{\varphi}(x) \right| \xrightarrow[n \to \infty]{} 0 \ .$$

Example 3.4

If f and φ are bounded and if s is periodic with period T then C is valid with

$$\overline{f}(\cdot) = \frac{1}{T} \sum_{t=1}^T f(\cdot - s_t)$$

and

$$\overline{\varphi}(\cdot) = \frac{1}{T} \sum_{t=1}^T \int yg(x - s_t, y - s_t)dy.$$

Example 3.5

If f satisfies a Lipschitz condition and if $s_n \to s$ (respectively s is bounded and $\frac{1}{n} \sum_{t=1}^n |s_t| \to 0$) then $\overline{f}(\cdot) = f(\cdot - s)$ (respectively $\overline{f} = f$). Furthermore, if φ satisfies a Lipschitz condition, if f is bounded and if s is bounded and $\frac{1}{n} \sum_{t=1}^n |s_t| \to 0$, then $\overline{\varphi} = \varphi$.

Example 3.6

A simple example of non periodic s should be

$$s_t = 1/[1 + \exp(-at + b)] \ , \ t > 0$$

with $a > 0$; it corresponds to a *logistic trend*.

Finally, note that the condition $\dfrac{1}{n} \sum\limits_{t=1}^{n} |s_t| \to 0$ may be compatible with the appearance of some *outliers*.

We now define a *pseudo-regression* setting

$$
\begin{aligned}
\overline{r}(x) &= \frac{\overline{\varphi}(x)}{\overline{f}(x)} \quad \text{if } \overline{f}(x) > 0 \\
&= E\xi_0 + \limsup \frac{1}{n} \sum_{t=1}^{n} s_t \quad \text{if } \overline{f}(x) = 0.
\end{aligned}
$$

$\overline{r}(x)$ appears as an approximation of $E(\eta_{n+1}|\eta_n = x)$. If for instance $s_n \to s$, we have for a continuous r and $\overline{f}(x) > 0$:

$$
\overline{r}(x) = s + r(x - s) = \lim_{n \to \infty} E(\eta_{n+1}|\eta_n = x).
$$

If s is periodic with period T, a rough estimate of

$$
|\,\overline{r}(x) - E(\eta_{n+1}|\eta_n = x)\,| \text{ is } |\frac{1}{T} \sum_{t=1}^{T} (s_{n+1} - s_t)|.
$$

The kind of robustness we deal with here consists in the fact that the kernel predictor

$$
(3.37) \qquad r_n(\eta_n) = \sum_{t=1}^{n-1} \eta_{t+1} K\left(\frac{\eta_n - \eta_t}{h_n}\right) \Big/ \sum_{t=1}^{n} K\left(\frac{\eta_n - \eta_t}{h_n}\right)
$$

is a good approximation of $\overline{r}(\eta_n)$. This property is specified in the following statement :

THEOREM 3.8 *If (ξ_t) satisfies the conditions of Theorem 3.5 and if C holds, then for some $\delta' \geq 0$*

$$
(3.38) \qquad n^{\delta'}[r_n(\eta_n) - \overline{r}(\eta_n)] \to 0 \quad a.s.
$$

besides $\delta' > 0$ if $\delta > 0$.

The proof is similar to that of Theorem 3.2 and therefore omitted. Details may be found in [BO].

Generalisation

One may consider the model

$$
(3.39) \qquad \eta_t = \xi_t + S_t \quad , \quad t \in \mathbb{Z} ,
$$

where (S_t) is a bounded process independent of (ξ_t). Then an analogous result may be established.

3.4.3 Related topics

We now briefly consider some extensions.

Interpolation

Let $(\xi_t, t \in \mathbb{Z})$ be a real strictly stationary process observed at times $-n_1, \ldots, -1, +1, \ldots n_2$. The interpolation problem consists in evaluating the missing data ξ_0.

Consider the associated process

$$Z_{t,n} = (X_{t,n}, Y_{t,n}) = [(\xi_{t-\ell_n}, \ldots, \xi_{t-1}, \xi_{t+1}, \ldots, \xi_{t+k_n}), \xi_t]$$

where $t \in E_n = \{-n_1 + \ell_n, \ldots, -k_n - 1, \ell_n + 1, \ldots, n_2 - k_n\}$ with $0 < \ell_n < n_1 = n_1(n)$, $0 < k_n < n_2 = n_2(n)$.

Making use of a strictly positive $\ell_n + k_n$-dimensional kernel K_n we construct the interpolator

$$(3.40) \qquad \hat{\xi}_{0,n} = \frac{\displaystyle\sum_{t \in E_n} Y_{t,n} K_n \left(\frac{X_{0,n} - X_{t,n}}{h_n} \right)}{\displaystyle\sum_{t \in E_n} K_n \left(\frac{X_{0,n} - X_{t,n}}{h_n} \right)}$$

which may be interpreted as an approximation of

$$E(\xi_0 \mid \xi_{-1}, \ldots, \xi_{-\ell_n}, \xi_1, \ldots, \xi_{k_n}).$$

Then, with slight modifications, *the results concerning the nonparametric predictor remain valid for* $\hat{\xi}_{0,n}$. For details we refer to [RH].

Obviously $\hat{\xi}_{0,n}$ may also be used for *detecting outliers* by comparing an observed random variable ξ_{t_0} with its interpolate $\hat{\xi}_{t_0,n}$. If we adopt the simple scheme (3.36) we obtain a test problem with null hypothesis $H_0 : s_{t_0} = 0$.

In order to construct a test we suppose that $t_0 = 0$ and under H_0 we set

$$p(\eta) = P(|\xi_0 - E(\xi_0 \mid X_{0,n})| > \eta) , \eta > 0$$

and

$$q(\varepsilon) = \inf\{\eta : p(\eta) \le \varepsilon\} , 0 < \varepsilon < 1.$$

Now a natural estimator for p is

$$p_n(\eta) = \frac{1}{\sharp E_n} \sum_{t \in E_n} \mathbf{1}_{|\xi_t - \hat{\xi}_{t_n}| > \eta}.$$

It induces an estimator for q defined by

$$q_n(\varepsilon) = \inf\{\eta : p_n(\eta) \leq \varepsilon\}$$

hence a critical region of the form

$$|\xi_0 - \hat{\xi}_{0,n}| > q_n(\varepsilon).$$

Chaos

Consider the dynamical system defined by (2.43) :

$$X_t = r(X_{t-1}) \ , \ t = 1, 2, \ldots$$

then, if $d = 1$, r is the regression of X_t on X_{t-1} and it can be estimated by the kernel method. Under classical conditions we have (cf. [MA])

$$E(r_n(x) - r(x))^2 = O\left(\left(\frac{\mathrm{Log}\, n}{n}\right)^{4/(d+4)}\right).$$

Note that $\hat{r}_n$ (defined by (2.57)) furnishes an alternative estimator for r.

Regression with error

The problem of regression with error may be stated as follows :

Let $\left(X_t^{(1)}, Y_t\right)$, $t \in \mathbb{Z}$ be a $\mathbb{R}^2$-valued strictly stationary process observed at times $1, \ldots, n$. $\left(X_t^{(1)}\right)$ has the decomposition

$$X_t^{(1)} = X_t + \varepsilon_t \ , \ t \in \mathbb{Z}$$

where the ε_t's are i.i.d; and where (X_t) and (ε_t) are independent.

The problem is to estimate $r(\cdot) = E(m(Y_t)|X_t = \cdot)$ where m is some real mesurable function such that $E|(m(Y_t)| < +\infty$.

In the particular case where ε_0 has a known density, say f_ε, the estimator takes the form

$$r_n(x) = \frac{\displaystyle\sum_{t=1}^{n} m(Y_t)\tilde{K}_n\left(\frac{x - X_t}{h_n}\right)}{\displaystyle\sum_{t=1}^{n} \tilde{K}_n\left(\frac{x - X_t}{h_n}\right)}, x \in \mathbb{R}$$

where $\tilde{K}_n$ is given by (2.60).

Now the asymptotic results are similar to those indicated in 2.5.3 : good convergence rates if ϕ_ε is algebraically decreasing and poor rates if ϕ_ε is geometrically decreasing. See [MS].

Note that this model is different from (3.39) since here the observed process is stationary.

3.5 Implementation of nonparametric method

In the current section we discuss the practical implementation of the kernel estimators and predictors.

3.5.1 Stabilization of variance

If the observed process, say (ζ_t), possesses a marked trend characterized by a non-constant variance, this one may be eliminated by using a preliminary transformation of the data.
For positive ζ_t 's an often used method is the so-called BOX and COX transformation defined as

$$T_\lambda(\zeta_t) = \frac{\zeta_t^\lambda - 1}{\lambda} \; , \lambda > 0$$

$$T_0(\zeta_t) = \mathrm{Log}\zeta_t = \lim_{\lambda \to 0(+)} T_\lambda(\zeta_t)$$

where λ has to be estimated (cf. [GE]).

If the variance of ζ_t is known to be proportional to the mean (respectively the square of the mean) then $\lambda = \frac{1}{2}$ (respectively $\lambda = 0$) is adequate.

If the variability of (ζ_t) is unknown one can estimate λ by minimizing

$$\sum_{t=1}^{n} \left(T_\lambda(\zeta_t) - \overline{\zeta}_n\right)^2$$

where $\overline{\zeta}_n = \dfrac{1}{n} \displaystyle\sum_{t=1}^{n} \zeta_t$.

3.5.2 Trend and seasonality

Let (η_t) be a real process with constant variance. It may be represented by the classical decomposition model

$$(3.41) \qquad \eta_t = \mu_t + \sigma_t + \xi_t \ , \ t \in \mathbb{Z}$$

where (μ_t) is a slowly varying function (the "trend component"), (σ_t) a periodic function with known period τ (the "seasonal component") and (ξ_t) a stationary zero mean process.

If μ and σ have a parametric form, their *estimation* may be performed using least square method. Suppose for instance that

$$(3.42) \qquad \mu_t = a_0 + a_1 t + \ldots + a_p t^p$$

and that

$$(3.43) \qquad \sigma_t = c_1 \sigma_{1t} + \ldots + c_\tau \sigma_{\tau t}$$

where

$$\sigma_{kt} = \mathbf{1}_{\{t=k(\bmod \tau)\}}; k = 1, \ldots, \tau.$$

Since $\sum_{k=1}^{\tau} \sigma_{kt} = 1$ it is necessary to introduce an additional condition which should ensure the identifiability of the model. A natural condition is

$$(3.44) \qquad \sum_{k=1}^{\tau} c_k = 0 \ ,$$

which expresses the compensation of seasonal effects over a period.

Now, given the data $\eta_1, \ldots, \eta_n$, the least square estimators of $a_1, \ldots, a_p, c_1, \ldots, c_\tau$ are obtained by minimizing

$$\sum_{t=1}^{n} (\eta_t - \mu_t - \sigma_t)^2$$

under the constraint (3.44).

The *elimination* of μ_t and σ_t is an alternative technique which seems preferable to estimation because it is more flexible :

In absence of seasonality the trend may be approximated by *smoothing* considering for instance the moving average

$$(3.45) \qquad \hat{\mu}_t = \frac{1}{2q+1} \sum_{j=-q}^{q} \eta_{t+j} \ , \ q+1 \le t \le n-q$$

and then eliminated by constructing the artificial data

$$\hat{\xi}_t = \eta_t - \hat{\mu}_t \quad, \quad q + 1 \leq t \leq n - q.$$

Another method of elimination is *differencing* :
Let us consider the first difference operator ∇ and its powers defined as

$$\nabla \eta_t = \eta_t - \eta_{t-1}$$

and

$$\nabla_k \eta_t = \nabla \left(\nabla^{k-1} \eta_t \right) \quad, \quad k \geq 1,$$

then if μ_t has the polynomial form (3.42) we get

(3.46) $$\nabla^p \eta_t = p! a_p + \nabla^p \xi_t \quad, \quad t \in \mathbb{Z}$$

and consequently $(\nabla^p \eta_t)$ is a stationary process with mean $p! a_p$.

In the general case where both trend and seasonality appear, the first step is to approximate the trend by using a moving average which eliminates the seasonality. If the period τ is even, one may set $q = \dfrac{\tau}{2}$ and put

(3.47) $$\mu_t^* = \frac{1}{q} \left(\frac{1}{2} \eta_{t-q} + \eta_{t-q+1} + \ldots + \eta_{t+q-1} + \frac{1}{2} \eta_{t+q} \right) \quad,$$

$q \leq t \leq n - q$. If τ is odd, one may use (3.45) with $q = \dfrac{\tau - 1}{2}$.
Now, in order to approximate the seasonal component one may consider

$$\hat{c}_k = v_k - \frac{1}{\tau} \sum_{j=1}^{\tau} v_j \quad; \quad k = 1, \ldots, \tau$$

where v_j denotes the average of the quantities $\eta_{j+i\tau} - \hat{\mu}_{j+i\tau}$, $q < j + i\tau \leq n - q$.

Then, considering the artificial data $\eta_t = \hat{c}_t$ (where $\hat{c}_t = \hat{c}_k$ if $t = k(\text{mod } \tau)$), one obtains a model with trend and without seasonality which allows the use of (3.45).

Some details about the above method may be found in [BD].

Note that differencing may also be used for seasonality. Here the difference operator is given by

$$\nabla_\tau \eta_t = \eta_t - \eta_{t-\tau}.$$

Applying ∇_τ one obtains the non-seasonal model.

$$\nabla_\tau \eta_t = (\mu_t - \mu_{t-\tau}) + (\xi_t - \xi_{t-\tau}).$$

Clearly all the above techniques suffer the drawback of perturbating the data. Thus, if $s_t = \mu_t + \sigma_t$ does not vary too much the model (3.36) may be considered.

In that case a "cynical" method consists in *ignoring s_t* ! The discussion and result in subsection 3.4.2 show that, in a nonparametric context, this method turns to be effective.

3.5.3   Construction of nonparametric estimators for stationary processes

If the observed process, say (ξ_t), is known to be stationary and if one wants to estimate the marginal density f or the regression $r(\cdot) = E(m(\xi_{t+H}) \mid (\xi_t, \ldots, \xi_{t-k+1}) = \cdot)$, the construction of a kernel estimator requires a choice of K and h_n.

Some theoretical results (cf. [EP]) show that the choice of K does not much influence the asymptotic behaviour of f_n or r_n : the naive kernel, the normal kernel and the Epanechnikov kernel are more or less equivalent.

On the contrary the choice of h_n turns to be crucial for the estimator's accuracy. A great deal of papers have been published on the subject. We refer to BERLINET-DEVROYE (1994) for a comprehensive treatment and an extensive bibliography concerning the density. For the regression one may consults the books by HÄRDLE ([HA-1] and [HA-2]). Here for the sake of simplicity we only discuss the problem for one-dimensional densities. The general case may be treated similarly.

a) **Plug-in method**
The best asymptotic choice of h_n at a point x is given by (2.7) : if $h_n = c_n n^{-1/5}$ where $c_n \to c > 0$ and if assumptions of Theorem 2.1 hold, then

$$n^{4/5} E(f_n(x) - f(x))^2 \;\; \to \;\; \frac{c^4}{4} f''^2(x) \left(\int u^2 K(u) du \right)^2$$

(3.48)

$$+ \frac{f(x)}{c} \int K^2$$

thus, the best c at x is

(3.49) $$c_0(x) = \left(\frac{f''^2(x)}{f(x)} \right)^{-1/5} \left(\frac{(\int u^2 K(u) du)^2}{\int K^2} \right)^{-1/5} .$$

Now, it may be easily proved that

$$
n^{4/5} E \parallel f_n - f \parallel_{L^2(\lambda)}^2 \quad \rightarrow \quad \frac{c^4}{4} \int f''^2 \left(\int u^2 K(u) du \right)^2
$$

(3.50)
$$
+ \frac{\int K^2}{c}
$$

thus, the best c associated with the asymptotic *Mean integrated square error* (MISE) is

$$
(3.51) \qquad c_0(f) = \left(\int f''^2 \right)^{-1/5} \left(\frac{(\int u^2 K(u) du)^2}{\int K^2} \right)^{-1/5} .
$$

The estimation of $c_0(x)$ and $c_0(f)$ requires the construction of preliminary estimates of f and f''.

For that purpose we may choose $K(u) = \dfrac{1}{\sqrt{2\pi}} e^{-\frac{u^2}{2}}$ and consider the case where $f(x) = \dfrac{1}{\sigma\sqrt{2\pi}} e^{-\frac{x^2}{2\sigma^2}}$. Then $c_0(f)$ may be approximated by $\hat{\sigma}_n = \left(\dfrac{1}{n} \Sigma(\xi_t - \bar{\xi}_n)^2 \right)^{1/2}$, and a convenient choice of h_n is

$$
(3.52) \qquad\qquad \hat{h}_n = \frac{\hat{\sigma}_n}{n^{1/5}}.
$$

An alternative choice of h_n should be the more robust

$$
(3.53) \qquad\qquad \widehat{h}_n = \frac{\xi_{([\frac{3n}{4}])} - \xi_{([\frac{n}{4}])}}{n^{1/5}},
$$

where $\xi_{(1)}, \dots, \xi_{(n)}$ denotes the order statistics associated with $\xi_1, \dots, \xi_n$.

The above considerations lead to the preliminary estimate

$$
(3.54) \qquad \hat{f}_n(x) = \frac{1}{n^{4/5} \hat{\sigma}_n \sqrt{2\pi}} \sum_{t=1}^{n} \exp\left(-\frac{n^{2/5}}{2\hat{\sigma}_n^2} (x - \xi_t)^2 \right), x \in \mathbb{R},
$$

and $\hat{f}_n''$ may be taken as an estimate of f''. Note that if the graph of $\hat{f}_n$ is too erratic it should be useful to smooth it by using polynomial interpolation before performing the derivation.

Now the final estimates f_n^* and f_n^{**} are constructed from $\hat{f}_n$ and $\hat{f}_n''$ by setting

(3.55) $h_n^*(x) = (2\sqrt{\pi})^{1/5} \left(\dfrac{\hat{f}_n''(x)}{\hat{f}_n(x)} \right)^{1/5} n^{-1/5} \ , x \in \mathbb{R}$

and

(3.56) $h_n^{**} = (2\sqrt{\pi})^{1/5} \left(\int \hat{f}_n''^2(x) \right)^{-1/5} n^{-1/5}$

hence

(3.57) $f_n^*(x) = \dfrac{1}{nh_n^*(x)} \sum_{t=1}^{n} \dfrac{1}{\sqrt{2\pi}} \exp\left(-\dfrac{1}{2} \left(\dfrac{x - \xi_t}{h_n^*(x)} \right)^2 \right) \ , x \in \mathbb{R}$

and

(3.58) $f_n^{**}(x) = \dfrac{1}{nh_n^{**}} \sum_{t=1}^{n} \dfrac{1}{\sqrt{2\pi}} \exp\left(-\dfrac{1}{2} \left(\dfrac{x - \xi_t}{h_n^{**}} \right)^2 \right) \ , x \in \mathbb{R}.$

b) **Cross-validation**
   If the regularity of f is unknown one can employ an empirical maximum likelihood method for the determination of h.

Let us suppose again that K is the normal kernel and consider the empirical likelihood

$$L(h) = \prod_{t=1}^{n} f_{n,h}(\xi_t) \ , \ h > 0$$

where

$$f_{n,h}(x) = \dfrac{1}{nh} \sum_{s=1}^{n} K\left(\dfrac{x - \xi_s}{h} \right) \ , x \in \mathbb{R}$$

we have $\sup_{h>0} L(h) = +\infty$ since

$$L(h) \geq \left(\dfrac{K(0)}{nh} \right)^n \xrightarrow[h \to 0]{} +\infty \ .$$

It is possible to remove that difficulty by setting

$$L_V(h) = \prod_{t=1}^{n} f_{n-1,h}^{(t)}(\xi_t)$$

where

$$f_{n-1,h}^{(t)}(\xi_t) = \dfrac{1}{(n-1)h} \sum_{s \neq t} K\left(\dfrac{x - X_s}{h} \right) .$$

We now have

$$\| f_{n-1,h}^{(t)} \|_\infty \leq \frac{K(0)}{h} \xrightarrow[h \to +\infty]{} 0$$

and

$$f_{n-1,h}^{(t)}(\xi_t) \xrightarrow[h \to 0]{} 0 .$$

Then the empirical maximum likelihood estimate $\tilde{h}_n$ does exist, hence the estimate

$$(3.59) \qquad \tilde{f}_n(x) = \frac{1}{n\tilde{h}_n} \sum_{t=1}^{n} \frac{1}{\sqrt{2\pi}} \exp \left(-\frac{1}{2} \left(\frac{x - \xi_t}{\tilde{h}_n} \right)^2 \right) , x \in \mathbb{R}.$$

Conclusion

Note first that other interesting methods are discussed in [BE-DE], particularly the double kernel method.

Now, the comparison between the various methods is somewhat difficult. It should be noticed that the normal kernel (or the EPANECHNIKOV kernel) and $\hat{h}_n = \hat{\sigma}_n n^{-1/5}$ are commonly used in practice and that they provide good results in many cases.

3.5.4 Prediction

If k is chosen or if (ξ_t) is known to be a kth order Markov process, the predictor is directly given by (3.19).

In the general case it is necessary to choose a suitable k (or k_n, see 3.4.1).

For convenience we suppose that (ξ_t) is a real process, m the identity and $H = 1$. Now let us consider the statistics

$$(3.60) \qquad \Delta_N(k) = \sum_{t=n_0}^{N} (\xi_t - \hat{\xi}_t(k))^2 , 1 \leq k \leq k_0$$

where n_0 and k_0 are given and where $\hat{\xi}_t(k)$ stands for the predictor of ξ_t based on the data $\xi_1, \ldots, \xi_{t-1}$ and associated with the regression $E(\xi_t | (\xi_{t-1}, \ldots, \xi_{t-k}) = \cdot)$. The minimization of $\Delta_N(k)$ gives a suitable k, say $\hat{k}_N$. Some variants of (3.60) are considered in [CA-DE] and in the appendix of this book.

Now a regression estimator $r_{N-\hat{k}_N}$ associated with $\hat{k}_N$ is constructed and using (3.19) we obtain finally the predictor $r_{N-\hat{k}_N}(\xi_N, \ldots, \xi_{N-\hat{k}_N+1})$. Note that the method is valid even if the process is not stationary, provided the data should be of the form (3.36). Otherwise one can use the methods indicated in

3.5.1 and 3.5.3.

Finally it is noteworthy that presence of exogeneous variables does not change the method since these ones can be integrated in the nonparametric model.

3.5.5   Comparison with parametric predictors

The popular BOX and JENKINS method is based on the ARMA(p, q) model. Recall that a real process $(\xi_t, t \in \mathbb{Z})$ is said to be ARMA(p, q) if it satisfies a relation of the form

$$(3.61) \qquad \xi_t - \phi_1 \xi_{t-1} - \ldots - \phi_p \xi_{t-p} = \varepsilon_t - \theta_1 \varepsilon_{t-1} - \ldots - \theta_q \varepsilon_{t-q}$$

where (ε_t) is a white noise (i.e. the ε_t's are i.i.d. and such that $0 < \sigma^2 = E\varepsilon_t^2 < +\infty$, $E\varepsilon_t = 0$) and $\phi_1, \ldots, \phi_p, \theta_1, \ldots, \theta_q$ are real parameters.

If the polynomials

$$\phi(z) = 1 - \phi_1 z - \ldots - \phi_p z^p$$

and

$$\theta(z) = 1 - \theta_1 z - \ldots - \theta_q z^q$$

have no common zeroes and are such that $\phi(z)\theta(z) \neq 0$ for $|z| \leq 1$ then (3.61) admits a unique stationary solution.

The BOX and JENKINS method mainly consists of the following steps :

1. Elimination of trend and seasonality by differencing.

2. Identification of (p, q).

3. Estimation of $\theta_1, \ldots, \theta_q, \phi_1, \ldots, \phi_p, \sigma^2$.

4. Construction of the predictor by using the estimated model.

For details we refer to [BO-JE], [G-M] and [BR-DA].

Now numerical results show that the nonparametric predictor is more precise than the [BO-JE] predictor. The most convincing comparative experience has been performed by CARBON and DELECROIX : they have considered 17 series constructed by simulation or taken from engineering, economics and physics and have found that the nonparametric predictor is better in at least 12 cases and equivalent in the other ones (see the appendix).

On the other hand a very complete work made by POGGI ([PO]), about prediction of global french electricity consumption, shows the great quality of nonparametric methods (cf the appendix).

Among other works let us indicate [DC-OP-TH] where the air pollution in Paris is studied by nonparametric methods (including exogeneous variables).

The quality of these predictors may be explained by the robustness pointed out in 3.4.2. In fact, the nonparametric method uses the information supplied by the history of the process (including seasonality) while the parametric technic needs to eliminate trend and seasonality before the construction of a stationary model.

Notes.

Estimation of regression function by the kernel method was first investigated by NADARAJA (1964) and WATSON (1964). A great number of people have studied the problem in the i.i.d. case. An early bibliography has been collected by COLLOMB (1981).

Theorem 3.1 is due to BOSQ and CHEZE (1994). Theorems 3.2, 3.3 are taken from BOSQ (1991) and RHOMARI (1994). Theorem 3.4 and results about prediction and interpolation for k^{th} Markov processes and general stationary processes are mainly due to RHOMARI (1994). For related results see the references.

Prediction for non stationary processes is taken from BOSQ (1991).

Concerning the implementation, (3.48) was suggested by DEHEUVELS and HOMINAL (1980). For thorough studies of the choice of h_n we refer to BERLINET and DEVROYE, BRONIATOWSKI, MARRON, HÄRDLE, SARDA, VIEU among others.

The practical comparison with parametric predictors has been performed by CARBON and DELECROIX (1992).

Chapter 4

Density estimation for continuous time processes

In this chapter we investigate the problem of estimating density for continuous time processes when continuous or sampled data are available.

We shall see that the situation is somewhat different from the discrete case. In fact, if the observed process paths are slowly varying the optimal rates are the same as in the discrete case. If, on the contrary, these paths are irregular one obtains *superoptimal rates* in quadratic mean and uniformly almost surely. It is noteworthy that these rates are preserved if the process is observed at judicious discrete instants.

In Section 1 we introduce the kernel estimator in a continuous time context. Section 2 is devoted to the quadratic error while Section 3 deals with uniform convergence. Sampling is considered in Section 4. Asymptotic normality appears in Chapter 5.

4.1 The kernel density estimator in continuous time

Let $(X_t, t \in \mathbb{R})$ be a $\mathbb{R}^d$-valued continuous time process defined on a probability space $(\Omega, \mathcal{A}, P)$. In all the following we assume that (X_t) is *measurable* (i.e. $(t, \omega) \to X_t(\omega)$ is $\mathcal{B}_{\mathbb{R}} \otimes \mathcal{A} - \mathcal{B}_{\mathbb{R}^d}$ measurable).

Suppose that the X_t's have a common distribution μ. We wish to estimate μ from the data $(X_t, 0 \leq t \leq T)$. A primary estimator for μ is the *empirical measure* μ_T defined as

(4.1)
$$\mu_T(B) = \frac{1}{T} \int_0^T \mathbf{1}_B(X_t)dt, \quad B \in \mathcal{B}_{\mathbb{R}^d}, \quad T > 0.$$

Now if μ has a density, say f, one may regularize μ_T by convolution, leading to the kernel density estimator defined as

$$(4.2) \qquad f_T(x) = \frac{1}{Th_T^d} \int_0^T K\left(\frac{x - X_t}{h_T}\right) dt, x \in \mathbb{R}^d$$

where K is a *kernel* (see Chapter 2) and where $h_T \to 0(+)$ as $T \to +\infty$.

In some situations we will consider the space $H_{k,\lambda}$ of the *kernels of order* (k, λ) $(k \in \mathbb{N}, 0 < \lambda \le 1)$ i.e. the space of mapping $K : \mathbb{R}^d \to \mathbb{R}$ bounded, integrable, such that $\int_{\mathbb{R}^d} K(u) du = 1$ and satisfying the conditions

$$\int_{\mathbb{R}^d} \| (u_1, \ldots, u_d) \|^\lambda |u_1|^{\alpha_1} \ldots |u_d|^{\alpha_d} |K(u_1, \ldots, u_d)| du_1 \ldots du_d < \infty$$

$$(4.3) \qquad\qquad\qquad and$$

$$\int_{\mathbb{R}^d} u_1^{\alpha_1} \ldots u_d^{\alpha_d} K(u_1, \ldots, u_d) du_1 \ldots du_d = 0,$$

$\alpha_1, \ldots, \alpha_d \in \mathbb{N}; \alpha_1 + \ldots + \alpha_d = j, 1 \le j \le k.$

Note that a *kernel* is a positive *kernel of order* $(1, 1)$. On the other hand we will use two mixing coefficients :

$$(4.4) \qquad \alpha^{(2)}(u) = \sup_{t \in \mathbb{R}} \sup_{\substack{A \in \sigma(X_t) \\ B \in \sigma(X_{t+u})}} |P(A \cap B) - P(A)P(B)|, \ u \ge 0$$

$$(4.5) \qquad \alpha(u) = \sup_{t \in \mathbb{R}} \sup_{\substack{A \in \sigma(X_s, s \le t) \\ B \in \sigma(X_s, s \ge t + u)}} |P(A \cap B) - P(A)P(B)|, \ u \ge 0$$

In particular we will say that (X_t) is GSM if $\alpha(u) \le a\rho^u, u > 0$ $(a > 0, 0 < \rho < 1)$.

Using the fact that $\sigma(X_s)$ is countably generated and employing the extension theorem (cf. [BI]) it is easy to check the measurability of $\alpha^{(2)}(\cdot)$. Similarly $\alpha(\cdot)$ is measurable as soon as (X_t) is CADLAG (i.e. the paths of (X_t) are continuous on the right and have a limit on the left at each t).

Concerning the properties of f we introduce the space $C_r^d(\ell)$ $(r = k + \lambda, 0 < \lambda \le 1, k \in \mathbb{N})$ of real valued function f, defined on $\mathbb{R}^d$, which are k times differentiable and such that

$$(4.6) \qquad \left| \frac{\partial f^{(k)}}{\partial x_1^{j_1} \ldots \partial x_d^{j_d}}(x') - \frac{\partial f^{(k)}}{\partial x_1^{j_1} \ldots \partial x_d^{j_d}}(x) \right| \le \ell \| x' - x \|^\lambda ;$$

$x, x' \in \mathbb{R}^d; j_1 + \ldots + j_d = k.$ Note that $C_{2,d}(b)$ is included in $C_2^d(b)$.

Finally it is interesting to note that the problem of estimating the finite dimensional distributions $P_{(X_{t_1}, \ldots, X_{t_m})}$ of (X_t) may be reduced to the problem of estimating μ by considering the $\mathbb{R}^{md}$-valued process

$$(4.7) \qquad Y_t = (X_t, X_{t+(t_2-t_1)}, \ldots, X_{t+(t_m-t_1)}) , \ t \in \mathbb{R}.$$

4.2 Optimal and superoptimal asymptotic quadratic error

In the current section we will assume that the X_t's have the same distribution but not a stronger condition like stationarity. We will see later the usefulness of that degree of freedom.

4.2.1 Consistency

Let us begin with a simple consistency result.

THEOREM 4.1 *If f is continuous at x and if $\alpha^{(2)} \in L^1(\lambda)$ then the condition $Th_T^{2d} \to +\infty$ implies*

$$(4.8) \qquad E(f_T(x) - f(x))^2 \to 0.$$

Furthermore if $f \in C_{k+\lambda}^d(\ell)$, $K \in H_{k,\lambda}$ and $h_T \simeq T^{-1/(2r+2d)}$ where $r = k + \lambda$ then

$$(4.9) \qquad E(f_T(x) - f(x))^2 = O(T^{-r/(r+d)}).$$

Proof

Using the classical Bochner's lemma (see (2.11)) we get

$$(4.10) \qquad E f_T(x) = \int_{\mathbb{R}^d} K_{h_T}(x - u) f(u) du \xrightarrow[h_T \to 0]{} f(x) .$$

Now Fubini's theorem entails

$$(4.11) \qquad \begin{aligned} V f_T(x) = \\ \frac{1}{T^2} \int_{[0,T]^2} \mathrm{Cov}\left(K_{h_T}(x - X_s), K_{h_T}(x - X_t)\right) ds dt, \end{aligned}$$

then by using covariance inequality (1.11) we obtain

$$V f_T(x) \leq \frac{4 \parallel K \parallel_\infty^2}{T^2 h_T^{2d}} \int_{[0,T]^2} \alpha^{(2)}(|t - s|) ds dt.$$

The integral on the right side, say I, may be written as

$$(4.12) \qquad \begin{aligned} I &= 2 \int_0^T dt \int_0^t \alpha^{(2)}(|t - s|) ds \\ &= 2 \int_0^T dt \int_0^t \alpha^{(2)}(u) du \leq 2T \int_0^{+\infty} \alpha^{(2)}(u) du \end{aligned}$$

and finally

$$(4.13) \qquad V f_T(x) \leq \frac{8 \parallel K \parallel_\infty^2}{T h_T^{2d}} \int_0^{+\infty} \alpha^{(2)}(u) du = O\left(1/T h_T^{2d}\right)$$

which leads to (4.8) by using (4.10) and

$$(4.14) \qquad E(f_T(x) - f(x))^2 = V f_T(x) + (E f_T(x) - f(x))^2.$$

Now, in order to prove (4.9) we study the bias of f_T.
Taylor formula and (4.3) entail

$$
\begin{aligned}
b_T(x) &= Ef_T(x) - f(x) = \int_{\mathbb{R}^d} K(u)[f(x - h_T u) - f(x)]du \\
&\equiv h_T^k \int_{\mathbb{R}^d} K(u) \sum_{j_1+\dots+j_d=k} \frac{u_1^{j_1} \dots u_d^{j_d}}{j_1! \dots j_d!} \frac{\partial^{(k)} f}{\partial x_1^{j_1} \dots \partial x_d^{j_d}}(x - \theta h_T u)du
\end{aligned}
$$

where $0 < \theta < 1$. Now using again (4.3) we obtain

$$
\begin{aligned}
b_T(x) = h_T^k \int_{\mathbb{R}^d} K(u) \sum_{j_1+\dots+j_d=k} \frac{u_1^{j_1} \dots u_d^{j_d}}{j_1! \dots j_d!} &\left[\frac{\partial^{(k)} f}{\partial u_1^{j_1} \dots \partial u_d^{j_d}}(x - \theta h_T u) \right. \\
&\left. - \frac{\partial^{(k)} f}{\partial u_1^{j_1} \dots \partial u_d^{j_d}}(x) \right] du
\end{aligned}
$$

and (4.6) implies

(4.15) $$|b_T(x)| \le c_{(r)} h_T^{k+\lambda}$$

where

(4.16) $$c_{(r)} = \sum_{j_1+\dots+j_d=k} \frac{\ell}{j_1! \dots j_d!} \int \| u \|^\lambda |u_1|^{j_1} \dots |u_d|^{j_d} |K(u)|du$$

thus (4.14), (4.13) and (4.15) yield

$$E(f_T(x) - f(x))^2 = O\left(\frac{1}{Th_T^{2d}}\right) + O\left(h_T^{2r}\right) = O\left(T^{-\frac{r}{r+d}}\right) . \blacksquare$$

4.2.2   Optimal rate

We now show that under mild mixing conditions the kernel estimator has at less the same rate, as in the i.i.d. case. This rate will be called "optimal rate".

In the following $g_{s,t} = f_{X_s,X_t} - f \otimes f$. We state the main assumptions

$A(\Gamma, p)$ - There exists $\Gamma \in \mathcal{B}_{\mathbb{R}^2}$ containing $D = \{(s,t) \in \mathbb{R}^2 : s = t\}$
  and $p \in]2, +\infty]$ such that

a) $g_{s,t}$ exists for $(s,t) \notin \Gamma$,

b) $\delta_p(\Gamma) = \sup_{(s,t)\notin\Gamma} \| g_{s,t} \|_{L^p(\mathbb{R}^{2d})} < +\infty$,

c) $\limsup_{T\to+\infty} \frac{1}{T} \int_{[0,T]^2 \cap \Gamma} dsdt = \ell_\Gamma < +\infty$.

$M(\gamma, \beta)$ - $\alpha^{(2)}(|t - s|) \le \gamma|t - s|^{-\beta}$; $(s,t) \notin \Gamma$
  where $\gamma > 0$ and $\beta > 0$.

The following lemma furnishes an upper bound for the variance of f_T.

LEMMA 4.1 *If $A(\Gamma, p)$ and $M(\gamma, \beta)$ hold for some Γ, p, γ, β, with $\beta \geq 2\dfrac{p-1}{p-2}$ then*

$$
(4.17) \quad
\begin{aligned}
V f_T(x) \leq{} & \frac{1}{Th_T^d} E\left[\frac{1}{h_T^d} K^2\left(\frac{x-X_0}{h_T}\right)\right] \cdot \frac{1}{T} \int_{[0,T]^2 \cap \Gamma} ds dt \\
& + \left(2 \parallel K \parallel_q^2 \delta_p(\Gamma) + \frac{8 \parallel K \parallel_\infty^2 \gamma}{\beta - 1}\right) \frac{h_T^\eta}{Th_T^d}
\end{aligned}
$$

where $q = \dfrac{p}{p-1}$ and $\eta = \dfrac{2d}{q}(1 - \dfrac{1}{\beta}) - d \geq 0$.

Proof

Let us consider the decomposition

$$
(4.18) \quad
\begin{aligned}
Th_T^d V f_T(x) ={} & \int_{[0,T]^2 \cap \Gamma} \mathrm{Cov}\left(K\left(\frac{x-X_s}{h_T}\right), K\left(\frac{x-X_t}{h_T}\right)\right) \frac{ds dt}{Th_T^d} \\
& + \int_{[0,T]^2 \cap \Gamma^c} \mathrm{Cov}\left(K\left(\frac{x-X_s}{h_T}\right), K\left(\frac{x-X_t}{h_T}\right)\right) \frac{ds dt}{Th_T^d}
\end{aligned}
$$

The first integral can be bounded above by

$$
(4.19) \quad I_T := \frac{1}{h_T^d} E K^2\left(\frac{x-X_0}{h_T}\right) \cdot \frac{1}{T} \int_{[0,T]^2 \cap \Gamma} ds dt.
$$

Concerning the second integral, we may use $A(\Gamma, p)$ and Holder's inequality with respect to Lebesgue measure for obtaining

$$
(4.20) \quad \left|\mathrm{Cov}\left(K\left(\frac{x-X_s}{h_T}\right), K\left(\frac{x-X_t}{h_T}\right)\right)\right| \leq h_T^{(2d)/q} \parallel K \parallel_q^2 \delta_p(\Gamma), \quad (s,t) \notin \Gamma ;
$$

thus, setting $E_T = \left\{(s,t) : |t-s| \leq h_T^{-(2d)/q\beta}\right\}$

we get

$$
\begin{aligned}
J_T : ={} & \frac{1}{Th_T^d} \int_{[0,T]^2 \cap \Gamma^c \cap E_T} \mathrm{Cov}\left[K\left(\frac{x-X_s}{h_T}\right), K\left(\frac{x-X_t}{h_T}\right)\right] ds dt \\
\leq{} & \frac{2}{Th_T^d} \int_{0 \leq s \leq t \leq T, t-s \leq h_T^{-(2d)/q\beta}} \parallel K \parallel_q^2 h_T^{(2d)/q} \delta_p(\Gamma) ds dt,
\end{aligned}
$$

hence

$$
(4.21) \quad J_T \leq 2 \parallel K \parallel_q^2 \delta_p(\Gamma) h_T^{((2d)/q)(1-\frac{1}{\beta})-d}.
$$

On the other hand, Billingsley's inequality (1.11) yields

$$
\begin{aligned}
J_T' : ={} & \frac{1}{Th_T^d} \int_{[0,T]^2 \cap \Gamma^c \cap E_T^c} \left|\mathrm{Cov}\left(K\left(\frac{x-X_s}{h_T}\right), K\left(\frac{x-X_t}{h_T}\right)\right)\right| ds dt \\
\leq{} & \frac{8 \parallel K \parallel_\infty^2}{Th_T^d} \int_{0 \leq s \leq t \leq T, t-s > h_T^{-(2d)/q\beta}} \gamma(t-s)^{-\beta} ds dt,
\end{aligned}
$$

hence

$$
(4.22) \quad J_T' \leq \frac{8 \parallel K \parallel_\infty^2 \gamma}{\beta - 1} h_T^{((2d)/q)\left(1-\frac{1}{\beta}\right)-d},
$$

and finally (4.19), (4.21) and (4.22) imply (4.17). ■

We are now in a position to state the result

THEOREM 4.2 *(Optimal rate).*

1) *If $A(\Gamma, p)$ and $M(\gamma, \beta)$ hold for some Γ, p, γ, β with $\beta > 2\dfrac{p-1}{p-2}$, f is continuous at x and $Th_T^d \to +\infty$ then*

$$(4.23) \qquad \limsup_{T\to\infty} Th_T^d V f_T(x) \le \ell_\Gamma f(x) \int K^2;$$

if f is bounded, then

$$(4.24) \qquad \limsup_{T\to\infty} \sup_{x\in\mathbb{R}^d} Th_T^d V f_T(x) \le \ell_\Gamma \parallel f \parallel_\infty \int K^2.$$

2) *If in addition $f \in C_r^d(\ell)(r = k+\lambda)$ and if $h_T = c_T T^{-1/(2r+d)}$ where $c_T \to c > 0$, then*

$$(4.25) \qquad \limsup_{T\to+\infty} \sup_{x\in\mathbb{R}^d} T^{2r/(2r+d)} E(f_T(x) - f(x))^2 \le C$$

where $C = \dfrac{\ell_\Gamma \parallel f \parallel_\infty \int K^2}{c^d}$

$$+c^{2r}\left(\sum_{j_1+...+j_d=k} \frac{\ell}{j_1!...j_d!} \int_{\mathbb{R}^d} \parallel u \parallel^\lambda |u_1|^{j_1} ... |u_d|^{j_d} K(u)du\right)^2.$$

Proof

1) Using (4.17) and noting that here η is strictly positive and that

$$\lim_{T\to\infty} EK_{h_T}^2(x - X_0) = f(x)\int K^2$$

we get (4.23). Concerning (4.24) it suffices to note that

$$EK_{h_T}^2(x - X_0) \le \parallel f \parallel_\infty \int K^2.$$

2) From (4.15) we deduce that

$$(4.26) \qquad \sup_{x\in\mathbb{R}^d} |Ef_T(x) - f(x)|^2 \le c_{(r)}^2 h_T^{2r}$$

where $c_{(r)}$ is given by (4.16). Thus (4.25) is a straightforward consequence of (4.24) and (4.26). ■

If $\beta = 2\dfrac{p-1}{p-2}$ (in particular if $p = +\infty$ and $\beta = 2$) the same rates are valid but with a constant greater than C.

In order to show that the above rates are achieved for some processes, let us consider the family $\mathcal{X}$ of processes $X = (X_t, t \in \mathbb{R})$ which satisfy the above hypothesis uniformly, in the following sense : there exist positive constants $f_0, L_0, \delta_0, \gamma_0, \beta_0$ and p_0 such that for each $X \in \mathcal{X}$ and with clear notations

- $\| f_X \|_\infty \le f_0$

- $\dfrac{1}{T} \displaystyle\int_{[0,T]^2 \cap \Gamma_X} ds\,dt \le L_0 \left(1 + \dfrac{1}{T}\right)$

- $p_X = p_0 > 2$ and $\delta_{p_0}(\Gamma_X) \le \delta_0$

- $\gamma \le \gamma_0$ and $\beta \ge \beta_0 > 2\dfrac{p_0 - 1}{p_0 - 2}$.

Then we have

COROLLARY 4.1

$$(4.27) \qquad \lim_{T \to +\infty} \max_{X \in \mathcal{X}} \sup_{x \in \mathbb{R}^d} Th_T^d V_X f_T(x) = L_0 f_0 \int K^2$$

where V_X denotes the variance if the underlying process is X.

Proof

An easy consequence of (4.17) is

$$(4.28) \qquad \limsup_{T \to +\infty} \max_{X \in \mathcal{X}} \sup_{x \in \mathbb{R}^d} Th_T^d V_X f_T(x) \le L_0 f_0 \int K^2.$$

It remains to exhibit a process $\widetilde{X}$ in $\mathcal{X}$ such that

$$(4.29) \qquad \sup_{x \in \mathbb{R}^d} Th_T^d V_X f_T(x) \xrightarrow[T \to \infty]{} L_0\, f_0 \int K^2.$$

To this aim we consider a sequence $(Y_n, n \in \mathbb{Z})$ of i.i.d. $\mathbb{R}^d$-valued random variables with a density f such that $\| f \|_\infty = f(x_0) = f_0$ for some x_0. Now let us set

$$(4.30) \qquad \widetilde{X}_t = Y_{[t/L_0]}, t \in \mathbb{R},$$

then $\widetilde{X}$ belongs to $\mathcal{X}$ with $\Gamma = \cup_{n \in \mathbb{Z}} \left\{ (s,t) : \left[\dfrac{s}{L_0}\right] = \left[\dfrac{t}{L_0}\right] = n \right\}$, $\ell_\Gamma = L_0$, $\delta_{p_0}(\Gamma) = 0$ and $\alpha^{(2)}(|t - s|) = 0$ if $(s,t) \in \Gamma^c$.

Now for that particular process f_T takes a special form, namely

$$(4.31) \qquad f_T(x) = \dfrac{[T/L_0]}{T/L_0} \hat{f}_T(x) + \dfrac{T - [T]}{Th_T^d} K\left(\dfrac{x - Y_{[T/L_0]}}{h_T}\right)$$

where $\hat{f}_T$ is a kernel estimator of f associated with the i.i.d. sample $Y_0, \ldots, Y_{[T/L_0]-1}$.

Then from (2.12) it is easy to deduce that

$$[T/L_0]h_T^d \sup_{x \in \mathbb{R}^d} V\hat{f}_T(x) \longrightarrow f_0 \int K^2$$

thus

$$Th_T^d \sup_{x \in \mathbb{R}^d} V \hat{f}_T(x) \longrightarrow L_0 f_0 \int K^2.$$

On the other hand

$$\sup_{x \in \mathbb{R}^d} V \left(\frac{T - [T]}{Th_T^d} K \left(\frac{x - Y_{[T/L_0]}}{h_T} \right) \right) \le \frac{\| K \|_\infty^2}{T^2 h_T^{2d}}$$

and finally

(4.32) $$Th_T^d \sup_{x \in \mathbb{R}^d} V_{\widetilde{X}} f_T(x) \longrightarrow L_0 f_0 \int K^2$$

which implies (4.27). ∎

It should be noticed that the process $\widetilde{X}$ is **not** stationary.

COROLLARY 4.2 *Let* $\mathcal{X}_1 = \{X : X \in \mathcal{X}, \ f_X \in C_r^d(\ell)\}$. *The choice* $h_T = c_T T^{-1/(2r+d)}$ *where* $c_T \to c > 0$ *implies*

(4.33) $$\limsup_{T \to +\infty} \sup_{X \in \mathcal{X}_1} \sup_{x \in \mathbb{R}^d} T^{2r/(2r+d)} E_X (f_T(x) - f(x))^2 \le C'$$

where $C' = \dfrac{L_0 f_0 \int K^2}{c^d} + c^{2r} c_{(r)}$.

Proof : Clear. ∎

The next theorem emphasizes the fact that the kernel estimator achieves the best convergence rate in a minimax sense.

THEOREM 4.3 *Let* $\mathcal{F}_T$ *be the class of all measurable estimators of the density based on the data* $(X_t, 0 \le t \le T)$ *then*

(4.34) $$\liminf_{T \to +\infty} \inf_{\tilde{f}_T \in \mathcal{F}_T} \sup_{X \in \mathcal{X}_1} T^{\frac{2r}{2r+d}} E_X \left(\tilde{f}_T(x) - f_X(x) \right)^2 > 0, \ x \in \mathbb{R}^d.$$

Proof (sketch)

Let $\mathcal{X}_0$ be the class of processes $X = (X_t, t \in \mathbb{R})$ $\mathbb{R}^d$-valued and such that

$$X_t = Y_{[t/L_0]}, \ \ t \in \mathbb{R}$$

where $(Y_n, n \in \mathbb{Z})$ is a sequence of i.i.d. r.v.'s with a density f belonging to $C_r^d(\ell)$ and such that $X \in \mathcal{X}$.

If $\tilde{f}_T \in \mathcal{F}_T$ then it induces an estimator $f_{[T]}$ which belongs to the family $\mathcal{F}_{[T]}^*$ of the measurable density estimators based on i.i.d. data $Y_1, \ldots, Y_{[T/L_0]}$.

Conversely each estimator $f_{[T]}^* \in \mathcal{F}_{[T]}^*$ generates $\tilde{f}_T \in \mathcal{F}_T$ by setting

$$\tilde{f}_T(x; X_t, 0 \le t \le T) = f_{[T]}^*(x; X_1, \ldots, X_{[T]}).$$

Now we clearly have

$$
\begin{aligned}
A_T := \inf_{\tilde{f}_T \in \mathcal{F}_T} \sup_{X \in \mathcal{X}} \; & T^{\frac{2r}{2r+d}} E_X \left(\tilde{f}_T(x) - f(x) \right)^2 \\
\geq \inf_{\tilde{f}_T \in \mathcal{F}_T} \sup_{X \in \mathcal{X}_1} \; & T^{\frac{2r}{2r+d}} E_X \left(\tilde{f}_T(x) - f(x) \right)^2 \\
= \inf_{f^*_{[T]}} \in \mathcal{F}^*_{[T]} \sup_{f \in C^d_r(\ell)} \; & T^{\frac{2r}{2r+d}} E \left(f^*_{[T]}(x) - f(x) \right)^2 =: B_T
\end{aligned}
$$

therefore

$\liminf_{T \to +\infty} A_T \geq \liminf_{T \to +\infty} B_T.$

Now applying a theorem of Ibragimov-Hasminski [IB-HA] (in fact a d-dimensional version of this theorem) we obtain

$$
\liminf_{T \to +\infty} B_T > 0
$$

hence (4.34). ∎

An easy adaptation of the above proof should provide an analogous minimax result for the discrete case.

Finally let us indicate that, like in the discrete case (see 2.2), similar results may be obtained replacing $A(\Gamma, p)$ by

$\Big[A'(\Gamma) - g_{s,t}$ exists for $(s,t) \notin \Gamma$ and is Lipschitzian uniformly with respect to (s,t), where Γ satisfies the condition

$$
\limsup_{T \to \infty} \frac{1}{T} \int_{[0,T]^2 \cap \Gamma} ds dt < +\infty.
$$

In that case the condition $\beta > 2(p-1)/(p-2)$ is replaced by the weaker condition $\beta > \dfrac{2d+1}{d+1}$.

4.2.3 Superoptimal rate

The following theorem produces a surprising result : if the distribution of (X_s, X_t) is not too close to a singular distribution for $|s - t|$ small then f_T converges at the "superoptimal rate" $\dfrac{1}{T}$.

Processes for which the rate T^{-1} is reached will be called "irregular paths processes".

THEOREM 4.4

1) If $g_{s,t} = g_{|t-s|}$ exists for $s \neq t$, if

$(y,z) \mapsto \displaystyle\int_{]0,+\infty[} |g_u(y,z)| du$ *is defined for any $(y,z) \in \mathbb{R}^{2d}$ is bounded and is continuous at (x,x) then*

$$
(4.35) \qquad \limsup_{T \to +\infty} T \cdot V f_T(x) \leq 2 \left(\int |K| \right)^2 \int_0^{+\infty} |g_u(x,x)| du
$$

2) If $g_{s,t} = g_{|t-s|}$ exists for $s \neq t$, if $u \rightarrow \parallel g_u \parallel_\infty$ is integrable on $]0, +\infty[$ and if g_u is continuous at (x, x) for each $u > 0$ then

$$(4.36) \qquad T \cdot V f_T(x) \rightarrow 2 \int_0^{+\infty} g_u(x, x) du.$$

Proof

1) Using (4.11) and the stationarity condition $g_{s,t} = g_{|t-s|}$ we get

$$(4.37) \qquad T \cdot V f_T(x) = 2 \int_0^T \left(1 - \frac{u}{T}\right) \mathrm{Cov}\left(K_{h_T}(x - X_0), K_{h_T}(x - X_u)\right) du .$$

Now for each $u > 0$

$$(4.38) \qquad \begin{aligned} &\mathrm{Cov}\left(K_{h_T}(x - X_0), K_{h_T}(x - X_u)\right) = \\ &\int_{\mathbb{R}^{2d}} K_{h_T}(x - y) K_{h_T}(x - z) g_u(y, z) dy dz \end{aligned}$$

therefore

$$(4.39) \qquad \begin{aligned} TV f_T(x) \leq \\ 2 \int |K_{h_T}(x - y) K_{h_T}(x - z)| \left(\int_0^{+\infty} |g_u(y, z)| du\right) dy dz \end{aligned}$$

taking lim sup on both side and applying Bochner's lemma we obtain (4.35).

2) Since $(u, y, z) \mapsto K_{h_T}(x - y) K_{h_T}(x - z) \left(1 - \frac{u}{T}\right) g_u(y, z) \mathbf{I}_{[0,T]}(u)$ is integrable we may apply Fubini's theorem to (4.37) and (4.38) leading to

$$(4.40) \qquad \begin{aligned} TV f_T(x) = \\ 2 \int_0^T K_{h_T}(x - y) K_{h_T}(x - z) \int_0^T \left(1 - \frac{u}{T}\right) g_u(y, z) du dy dz. \end{aligned}$$

Now

$$\left| \int_0^{+\infty} g_u(y, z) du - \int_0^T \left(1 - \frac{u}{T}\right) g_u(y, z) du \right| = \left| \int_T^\infty g_u - \int_0^T \frac{u}{T} g_u \right|$$

$$\leq \int_T^\infty \parallel g_u \parallel_\infty du + \int_0^T \frac{u}{T} \parallel g_u \parallel_\infty du; (y, z) \in \mathbb{R}^{2d};$$ then, the integrability of $\parallel g_u \parallel_\infty$ and the dominated convergence theorem show that the bound vanishes as $T \rightarrow +\infty$.
Hence

$$(4.41) \qquad \begin{aligned} TV f_T(x) = \\ 2 \int K_{h_T}(x - y) K_{h_T}(x - z) \left(\int_0^\infty g_u(y, z) du\right) dy dz + o(1). \end{aligned}$$

Now the dominated convergence theorem entails that $(y, z) \mapsto \int_0^{+\infty} g_u(y, z) du$ is continuous at (x, x) and finally Bochner's lemma implies (4.36). ■

COROLLARY 4.3

1) *If assumptions of Theorem 4.1 hold for each x,*
if $G = \sup_{x \in \mathbb{R}^d} \int_0^{+\infty} |g_u(x,x)| du < +\infty$ and if $f \in C_r^d(\ell)$ $(r = k + \lambda)$ and
$K \in H_{k,\lambda}$, then the choice $h_T = c_T T^{-1/(2r)}(c_T \mapsto c > 0)$ leads to

(4.42)
$$\limsup_{T \to +\infty} \sup_{x \in \mathbb{R}^d} TE(f_T(x) - f(x))^2 \le 2\left(\int |K|\right)^2 G + c^{2r} c_{(r)}^2.$$

2) *If assumptions of Theorem 4.2 hold, if $f \in C_{2,d}(b)$ if K is a (positive) kernel*
and if $h_T = c_T T^{-1/4}(c_T \to c > 0)$ then

(4.43)
$$TE(f_T(x) - f(x))^2 \xrightarrow[T \to \infty]{} 2 \int_0^{+\infty} g_u(x,x) du + \frac{c^4}{4} \chi_f^2(x)$$

where $\displaystyle \chi_f(x) = \sum_{i \le i, j \le d} \frac{\partial^2 f}{\partial x_i, \partial x_j}(x) \int u_i u_j K(u) du.$

Proof : Clear. ∎

Example 4.1
    Let $(X_t, t \in \mathbb{R})$ be a Gaussian real stationary process with zero mean and auto-correlation function
(4.44)
$$\rho(u) = 1 - a|u|^\theta + o(u^\theta)$$
when $u \to 0$, where $0 < \theta < 2$.

    Then it is easy to verify that

(4.45)
$$|g_u(x,y)| \le a|\rho(u)|\mathbf{I}_{|u|>b} + \left(c + d|u|^{-\frac{\theta}{2}}\right)\mathbf{I}_{|u| \le b, u \ne 0}$$

where a, b, c, d are suitable constants.

    Consequently, conditions in Corollary 4.3 are satisfied as soon as ρ is integrable on $]0, +\infty[$.

Example 4.2
    Let $(X_t, t \ge 0)$ be a real diffusion process defined by the stochastic differential equation
(4.46)
$$dX_t = S(X_t)dt + \sigma(X_t)dW_t, \ X_0 = x_0, \ t \ge 0$$
where S and σ satisfy a Lipschitz condition and the condition
$$I = \int_{\mathbb{R}} \sigma^{-2}(x) \exp\left\{2 \int_0^x S(y)\sigma^{-2}(y)dy\right\} dx < +\infty$$
and where $(W_t, t \ge 0)$ is a standard Wiener process.

    It may be proved that such a process admits a stationary distribution with density given by
$$f(x) = I^{-1}\sigma^{-2}(x) \exp\left\{2 \int_0^x S(y)\sigma^{-2}(y)dy\right\}, x \in \mathbb{R}.$$

Moreover, under some regularity assumptions on S and σ, the kernel estimator of f reaches the full rate $\frac{1}{T}$. In particular if X_0 has the density f, conditions of Corollary 4.3 are fulfilled (see [KU] and [LE]).

Y. KUTOYANTS has recently shown that $\frac{1}{T}$ is the **minimax rate** in the model defined by (4.46) (see [KU]).

4.2.4   Intermediate rates

It is natural to formulate the following problem : what are all the possible rates for density estimators in continuous time? We give a partial answer in the present subsection.

We begin with a proposition which shows that, in some sense, conditions in Theorem 4.4 are necessary for obtaining the superoptimal rate $\frac{1}{T}$.

THEOREM 4.5 *Let* $(X_t, t \in \mathbb{R})$ *be a* $\mathbb{R}^d$-*valued process such that*

(a) $g_{s,t} = g_{|t-s|}$ *exists for* $s \neq t$ *and* $\displaystyle\int_{u_0}^{+\infty} \parallel g_u \parallel_\infty du < \infty$ *for* $u_0 > 0$.

(b) f *is continuous at* x *and* $f_{(X_0, X_u)}$ *is continuous at* (x, x) *for* $u > 0$.

(c) $\displaystyle\int_0^{u_1} f_{(X_0, X_u)}(x, x) du = +\infty$ *for* $u_1 > 0$.

Then if K *is a strictly positive kernel*

(4.47) $$\lim_{T \to \infty} T V f_T(x) = +\infty\ ,$$

and consequently $T E (f_T(x) - f(x))^2 \to +\infty$.

Proof

We first consider the integral

$$I_T = 2 \int_{\mathbb{R}^{2d}} K_{h_T}(x - y) K_{h_T}(x - z) \int_{u_0}^{T} \left(1 - \frac{u}{T}\right) g_u(y, z) du dy dz\ ,\ T > u_0.$$

Using (a) we obtain the bound

$$|I_T| \leq 2 \int_{u_0}^{+\infty} \parallel g_u \parallel_\infty du.$$

On the other hand, (b) implies

$$\lim_{h_T \to 0} \int_{\mathbb{R}^{2d}} K_{h_T}(x - y) K_{h_T}(x - z) f(y) f(z) dy dz = f^2(x).$$

Then, by using (4.40) we get

$$TV f_T(x) =$$

$$2 \int_0^{u_0} \left(1 - \frac{u}{T}\right) du \int_{\mathbb{R}^{2d}} K_{h_T}(x - y) K_{h_T}(x - z) f_{(X_0, X_u)}(y, z) dy dz + O(1).$$

Now, since $T \geq 2u_0$ implies $2\left(1 - \dfrac{u_0}{T}\right) \geq 1$ it suffices to show that $J_T \to \infty$ where

$$J_T = \int_0^{u_0} du \int_{\mathbb{R}^{2d}} K_{h_T}(x - y) K_{h_T}(x - z) f_{(X_0, X_u)}(y, z) dy dz.$$

Since the integrand is positive we may apply Fubini's theorem for obtaining

$$J_T = \int_{\mathbb{R}^{2d}} K_{h_T}(x - y) K_{h_T}(x - z) dy dz \int_0^{u_0} f_{(X_0, X_u)}(y, z) du$$

where the inner integral is finite for λ^{2d}-almost all (y, z).

Now by considering the affine transformation $(y, z) \mapsto (x - h_T v, x - h_T w)$ and by using the image measure theorem (see [RA]) we obtain

$$J_T = \int_{\mathbb{R}^{2s}} K(v) K(w) dv dw \int_0^{u_0} f_{(X_0, X_u)}(x - h_T v, x - h_T w) du.$$

We are now in a position to conclude : Firstly (b), (c) and Fatou's lemma imply

$$\lim_{T \to \infty} \int_0^{u_0} f_{(X_0, X_u)}(x - h_T v, x - h_T w) du = +\infty$$

for λ^{2d} almost all (v, w).

Applying again Fatou's lemma we get $J_T \to \infty$ thus

$$TV f_T(x) \to \infty$$

and the proof is therefore complete. ∎

In the Gaussian case we have the following

COROLLARY 4.4 *Let $(X_t, t \in \mathbb{R})$ be a real stationary Gaussian process, continuous in mean square and such that*

(a) $|\mathrm{Cov}(X_0, X_u)| < V X_0, \quad u > 0$ *and*

$$\int_{u_0}^{+\infty} |\mathrm{Cov}(X_0, X_u)| du < \infty, u_0 > 0.$$

Then if K is a strictly positive kernel we have

(4.48)
$$\begin{cases} \int_0^{u_1} (E|X_u - X_0|^2)^{-1/2} du = +\infty \,, u_1 > 0 \Rightarrow T.V f_T(x) \to +\infty, \\ \int_0^{u_1} (E|X_u - X_0|^2)^{-1/2} du < +\infty \,, u_1 > 0 \Rightarrow T.V f_T(x) \to \ell < +\infty. \end{cases}$$

In particular if (X_t) has differentiable sample paths then
$$T V f_T(x) \to +\infty.$$

We see that, at least for Gaussian processes, the "full rate" is closely linked with the irregularity of sample paths. It is interesting to note that, in order to reach the full rate, continuity of (X_t) is **not** required.
For example if the autocorrelation satisfies

(4.49) $$1 - \rho(u) \cong_{u \to o(+)} \frac{1}{|\text{Log}(u)|^{1-\beta}}, \quad 0 < \beta < 1$$

(X_t) is not a.s. continuous (see [AD]) but $V f_T(x) \cong \frac{1}{T}$ provided (a) is satisfied.

Finally note that, using Theorem 4.2 one can construct an estimator such that $T^{1-\varepsilon} E(f_T(x) - f(x))^2 \to 0$ ($\varepsilon > 0$) a soon as the Gaussian process (X_t) satisfies mild mixing conditions. In particular D. BLANKE has recently proved that if ρ is monotonous in a neighbourhood of 0 and satisfies the condition $1 - \rho(u) \sim u^2$ as $u \to 0$ then

$$E(f_T(x) - f(x))^2 \simeq \frac{\text{Log}T}{T} .$$

Proof of Corollary 4.4
We may and do suppose that $EX_0 = 0$ and $EX_0^2 = 1$ and we put $\rho(u) = E(X_0 X_u)$.

Let us set $\varphi(u) = \left(1 - \rho^2(u)\right)^{-1/2}$, $u > 0$ then we have

$$f_{(X_0, X_u)}(y, z) = \frac{\varphi(u)}{2\pi} \exp\left(-\frac{\varphi(u)^2}{2}(y^2 - 2\rho(u)yz + z^2)\right) ;$$

$(y, z) \in \mathbb{R}^{2d}$, $u > 0$.

Here condition (c) in Theorem 4.5 may be written

$$\int_0^{u_1} \frac{\varphi(u)}{2\pi} \exp\left(-\frac{x^2}{1 + \rho(u)}\right) du = +\infty$$

which is equivalent to $\int_0^{u_1} \varphi(u)du = +\infty$ since $\lim_{u \to 0} \rho(u) = 1$ by mean square continuity. Thus we have clearly the first implication.

Now it is easy to check that $\int_0^{u_1} \varphi(u)du < \infty$ implies $\int_0^{u_1} \| g_u \|_\infty du < \infty$.
Then Theorem 4.4 entails $TV f_T(x) \to 2 \int_0^\infty g_u(x, x)du < \infty$, hence the second implication

Finally, if (X_t) has differentiable sample paths, they are differentiable in mean square too (see [IB]-[HA]) and consequently

$$E \left(\frac{X_u - X_0}{u}\right)^2 \xrightarrow[u \to 0]{} EX_0'^2 .$$

Condition (a) implies $EX_0'^2 > 0$, then

$$u\left(E(X_u - X_0)^2\right)^{-1/2} \to 1/\sqrt{EX_0'^2} > 0$$

which implies $\int_0^{u_1}\left(E(X_u - X_0)^2\right)^{-1/2} du = +\infty$ and therefore $T \cdot Vf_T(x) \to +\infty$. ∎

We now give sufficient conditions for rates between $\dfrac{1}{T^{r/r+d}}$ and $\dfrac{1}{T}$. We will use conditions $A'(p)$ where $p \in [1, +\infty]$ defined by
$A'(p)$ - $g_{s,t}$ exists for $s \neq t$, $\| g_{s,t} \|_p$ is locally integrable and

$$\limsup_{T \to +\infty} \frac{1}{T} \int_{]0,T]^2} \| g_{s,t} \|_p \, dsdt = G_p < +\infty.$$

Note that if $g_{s,t} = g_{|t-s|}$ we have

$$\frac{1}{T} \int_{]0,T]^2} \| g_{s,t} \|_p \, dsdt = \frac{1}{T} \int_0^T \left(1 - \frac{u}{T}\right) \| g_u \|_p \, du$$

so that $\| g_u \|_p$ is integrable over $]0, +\infty[$. Then $A'(p)$ is fulfilled with $G_p = 2\int_0^{+\infty} \| g_u \|_p \, du$. In particular assumptions in Theorem 4.2 imply $A'(+\infty)$. On the other hand if $\int_{u_0}^{+\infty} \| g_u \|_1 \, du < +\infty$ for some $u_0 > 0$, $A'(1)$ is satisfied since $\| g_u \|_1 \leq 2$.

We now state a result which links the convergence rate with $A'(p)$.

THEOREM 4.6

1) If $A'(p)$ holds for some $p \in [1, +\infty]$ then

(4.50)
$$\limsup_{T \to \infty} T h_T^{(2d)/p} V f_T(x) \leq G_p \| K \|_q^2$$

where $q = p/(p-1)$.

2) If in addition $f \in C_r^d(\ell)$ $(r = k + \lambda)$, $K \in H_{k,\lambda}$ and $h_T = c_T T^{-p/(2pr+2d)}$ $(c_T \to c > 0)$ then

(4.51)
$$\limsup_{T \to \infty} T^{pr/(pr+d)} E(f_T(x) - f(x))^2$$
$$\leq c^{-\frac{2d}{p}} G_p \| K \|_q^2 + c^{2r} c_{(r)}^2.$$

Proof

1) We have

$$Vf_T(x) = \frac{1}{T^2 h_T^{2d}} \int_{[0,T]^2} dsdt \left[\int_{\mathbb{R}^{2d}} K\left(\frac{x-u}{h_T}\right) K\left(\frac{x-v}{h_T}\right) g_{s,t}(u,v) dudv \right].$$

Applying Holder inequality in the inner integral we get

$$V f_T(x) \leq$$

$$\leq \frac{1}{T^2 h_T^{2d}} \left(\int_{\mathbb{R}^{2d}} K^q \left(\frac{x-u}{h_T} \right) K^q \left(\frac{x-v}{h_T} \right) du dv \right)^{1/q} \int_{[0,T]^2} \parallel g_{s,t} \parallel_p ds dt$$

$$\leq \frac{1}{T} \frac{\parallel K \parallel_q^2}{h_T^{2d-(2d)/q}} \frac{1}{T} \int_{[0,T]^2} \parallel g_{s,t} \parallel_p ds dt$$

hence (4.50).

2) Clear. ∎

Note that the optimal rate is reached for $p = 2$ and the parametric rate for $p = +\infty$. If $p = 1$ one obtains the same rate as in Theorem 4.1. Note however that each of these rate is not necessarily the best one when $A'(p)$ holds.

We complete this section with an example which shows that if the observed process is nonstationary any rate is possible. Consider the process

$$(4.52) \qquad X_t = Y_k \cos \left(\frac{\pi}{2} \frac{t - k^\gamma}{(k+1)^\gamma - k^\gamma} \right) + Y_{k+1} \sin \left(\frac{\pi}{2} \frac{t - k^\gamma}{(k+1)^\gamma - k^\gamma} \right) ;$$

$k^\gamma \leq t < (k+1)^\gamma$, $k \in \mathbb{Z}$; where γ is a strictly positive constant and where $(Y_k, k \in \mathbb{Z})$ is a sequence of i.i.d. real Gaussian zero mean r.v.'s with variance $\sigma^2 > 0$. The observation of (X_t) over $[0, T]$ is in fact equivalent to the observation of $Y_0, \ldots, Y_{[T^{1/\gamma}]}$ and the best rate is $T^{-1/\gamma}$ since an asymptotically optimal estimator is

$$\overline{f}_T(x) = \frac{1}{S_T \sqrt{2\pi}} \exp \left(-\frac{1}{2} \frac{x^2}{S_T^2} \right), x \in \mathbb{R}$$

where $S_T = \frac{1}{[T^{1/\gamma}] + 1} \sum_{j=0}^{[T^{1/\gamma}]} X_j$.

Note that the kernel estimator remains competitive since here r may be chosen arbitrarily large.

Finally, γ being any strictly positive number, we have a family of processes for which any rate of the form $T^{-1/\gamma}$ is attained.

4.3   Optimal and superoptimal uniform convergence rates

For the study of uniform convergence we need a **Borel-Cantelli type lemma for continuous time processes** :

LEMMA 4.2 Let $(Z_t, t \geq 0)$ be a real continuous time process such that

(a) *For each $\eta > 0$, there exists a real decreasing function φ_η, integrable on $\mathbb{R}_+$ and satisfying*

$$P(|Z_t| > \eta) \leq \varphi_\eta(t) \ , t > 0 \ ,$$

(b) *The sample paths of (Z_t) are uniformly continuous with probability 1. Then*

$$\lim_{T \to +\infty} Z_T = 0 \ a.s..$$

Proof

First let (T_n) be a sequence of real numbers which satisfies $T_{n+1} - T_n \geq a > 0$, $n \geq 1$ where a is some constant.

Since φ_η is decreasing we have

$$\int_{T_N}^{+\infty} \varphi_\eta(t)dt \geq \sum_{n \geq N}(T_{n+1} - T_n)\varphi_\eta(T_{n+1}) \geq a \sum_{n \geq N} \varphi_\eta(T_{n+1})$$

thus $\sum_n \varphi_\eta(T_n) < +\infty$ and the classical Borel-Cantelli lemma yields

$$P\left(\limsup_n \{|Z_{T_n}| > \eta\}\right) = 0, \eta > 0 \text{ which in turn implies } Z_{T_n} \to 0 \text{ a.s.}$$

Let now (T_n) be any sequence of real numbers satisfying $T_n \uparrow +\infty$.

To each positive integer k we may associate a subsequence $\left(T_p^{(k)}\right)$ of (T_n) defined as follows :

$$\begin{aligned} T_1^{(k)} &= T_{n_1} \quad \text{where} \quad n_1 = 1, \\ T_2^{(k)} &= T_{n_2} \quad \text{where} \quad T_{n_2} - T_{n_1} \geq \tfrac{1}{k}, T_{n_2} - T_{n_2-1} < \tfrac{1}{k}, \\ &\vdots \\ T_p^{(k)} &= T_{n_p} \quad \text{where} \quad T_{n_p} - T_{n_{p-1}} \geq \tfrac{1}{k}, T_{n_p} - T_{n_p-1} < \tfrac{1}{k}, \\ &\vdots \end{aligned}$$

The first part of the current proof shows that $Z_{T_p^{(k)}} \underset{p \to \infty}{\longrightarrow} 0$ a.s. for each k. Now let us set

$$\Omega_0 = \{\omega : t \mapsto Z_t(\omega) \text{ is uniformly continuous, } Z_{T_p^{(k)}} \to 0, k \geq 1\}$$

clearly $P(\Omega_0) = 1$.

Now if $\omega \in \Omega_0$ and $\eta > 0$ there exists $k = k(\eta, \omega)$ such that $|t - s| \leq \dfrac{1}{k}$ implies $|Z_t(\omega) - Z_s(\omega)| < \dfrac{\eta}{2}$. Consider the sequence $\left(T_p^{(k)}\right)$: for each p and each n such that $n_p \leq n < n_{p+1}$ we have $\left|T_n - T_{n_p}\right| < \dfrac{1}{k}$, hence $\left|Z_{T_n}(\omega) - Z_{T_{n_p}}(\omega)\right| < \dfrac{\eta}{2}$.

Now for p large enough we have $\left|Z_{T_{n_p}}(\omega)\right| < \frac{\eta}{2}$ and consequently $|Z_{T_n}(\omega)| < \eta$ for n large enough. This is valid for each $\eta > 0$ and each $\omega \in \Omega_0$, thus $Z_{T_n} \to 0$ a.s.

∎

4.3.1 Optimal rates

We make the following assumptions

- $A(\Gamma, p)$ holds for some (Γ, p),
- f is bounded and belongs to $C_r^d(\ell)$, $r = k + \lambda$,
- $K \in H_{k,\lambda}$ and $K = K_0^{\otimes d}$ where K_0 has compact support and continuous derivative,
- $h_T = c_T \left(\dfrac{\log T}{T}\right)^{\frac{1}{2r+d}}$ $(c_T \to c > 0)$.

We first derive upper bounds for $P(|Z_T| > \eta)$ where

$$Z_T = \frac{1}{\log_m T}\left(\frac{T}{\log T}\right)^{\frac{r}{2r+d}}(f_T(x) - Ef_T(x)).$$

LEMMA 4.3

1) If $\alpha(u) \leq \gamma u^{-\beta}$, $u > 0$ where $\beta > \max\left(2\dfrac{p-1}{p-2}, \dfrac{7r+5d}{2r}\right)$

   then

   (4.53) $P(|Z_T| > \eta) \leq \dfrac{A}{T^{1+\mu}}$, $\eta > 0, T \geq 1$

   where A and μ do not depend on x.

2) If (X_t) is GSM then

   (4.54) $P(|Z_T| > \eta) \leq \dfrac{B}{T^{C(\log_m T)^2}}$, $\eta > 0, T \geq 1$

   where B and C do not depend on x.

Proof

We may and do suppose that $c_T = 1$ and $\eta < 1$.

1) Let us set

(4.55) $Y_{jn} = \dfrac{1}{\delta}\displaystyle\int_{(j-1)\delta}^{j\delta} K_{h_T}(x - X_t)dt$; $j = 1, \ldots, n$

where $n\delta = T$, $n = [T]$ $(T \geq 1)$ and consequently $2 > \delta \geq 1$. Thus we have

(4.56) $f_T(x) - Ef_T(x) = \dfrac{1}{n}\displaystyle\sum_{j=1}^{n}(Y_{jn} - EY_{jn}).$

In order to apply inequality (1.26) we have to study $V\left(\displaystyle\sum_{j=1}^{p} Y_{jn}\right)$. To this aim we may use inequality (4.17) in Lemma 4.1 with $p\delta$ instead of T and p' instead of p

for convenience.
We have readily

$$V\left(\frac{1}{p\delta}\int_0^{p\delta} K_{h_T}(x - X_t)dt\right) \le \frac{1}{p\delta h_T^d}E\left[\frac{1}{h_T^d}K^2\left(\frac{x - X_0}{h_T}\right)\right]$$

(4.57)

$$\cdot\frac{1}{p\delta}\int_{[0,p\delta]^2\cap\Gamma} dsdt + \left(2\parallel K\parallel_{q'}^2 \delta_{p'}(\Gamma) + \frac{8\parallel K\parallel_\infty^2 \gamma}{\beta - 1}\right)\frac{h_T^{\frac{2d}{q'}\left(1-\frac{1}{\beta}\right)-d}}{p\delta h_T^d}$$

where $q' = p'/(p' - 1)$.
Therefore, since $\beta > 2\dfrac{p' - 1}{p' - 2}$ we have

$$V\left(\frac{1}{p\delta}\int_0^{p\delta} K_{h_T}(x - X_t)dt\right) \le \frac{a}{p\delta h_T^d}$$

where $a = a(K, \parallel f \parallel_\infty, d, \gamma, \beta)$ does not depend on x.

Consequently

(4.58)
$$V\left(\sum_{j=1}^p Y_{jn}\right) \le a\frac{p}{\delta h_T^d}$$

then noting that $|Y_{jn} - EY_{jn}| \le 2\parallel K\parallel_\infty h_T^{-d}$ we obtain

$$v^2(q) \le \frac{2a}{p\delta h_T^d} + \frac{\parallel K\parallel_\infty \varepsilon}{h_T^d} \ , \ \varepsilon > 0.$$

Now choosing $p = [\varepsilon^{-1}\delta^{-1}]$ we get

$$v^2(q) \le A_0\frac{\varepsilon}{h_T^d}$$

where A_0 is a positive constant.

Therefore, substituying in (1.26) we arrive at

$$P(|f_T(x) - Ef_T(x)| > \varepsilon) \le 4\exp\left(-\frac{1}{8A_0}\varepsilon qh_T^d\right)$$

(4.59)

$$+22\left(1 + \frac{8\parallel K\parallel_\infty}{\varepsilon h_T^d}\right)^{1/2} q\alpha(p) =: u_T + v_T.$$

Now we choose $\varepsilon = \varepsilon_T = h_T^r(\log_m T)\eta$ $(\eta > 0)$ and we notice that $q = \dfrac{n}{2p} \ge \dfrac{\varepsilon n\delta}{2} = \dfrac{\varepsilon T}{2}$, hence

$$u_T \le 4\exp\left(-\frac{1}{16A_0}\log T \cdot (\log_m T)^2\eta^2\right)$$

thus

(4.60)
$$u_T \le \frac{4}{T^{A\eta^2(\log_m T)^2}} \ , \ (A > 0).$$

We now turn to the study of v_T. Using the elementary inequality $(1+w)^{1/2} < 1 + w^{1/2}$ we get

(4.61) $$v_T \le c_1 q \alpha(p) + c_2 \varepsilon^{-1/2} h_T^{-d/2} q \alpha(p)$$

then, after some easy calculations, we obtain

$$v_T \le c_3 \eta^{\frac{1}{2}+\beta} T^{-\frac{2r\beta-3r-3d}{4r+2d}} (\log T)^{\frac{r+\beta-d}{4r+2d}} (\log_m T)^{1/2+\beta},$$

and since $\beta > \dfrac{7r+5d}{2r}$ we have the bound

(4.62) $$v_T \le \frac{c_4}{T^{1+\mu}} \qquad (c_4 > 0, \mu > 0).$$

If $\eta > 1$ it is easy to see that c_4 must be replaced by $c_4 \eta^{1+\beta}$. Collecting (4.60) and (4.62) we arrive at (4.53).

2) If $\alpha(\cdot)$ tends to zero at an exponential rate (4.60) remains valid but (4.62) may be improved. From (4.61) we derive the bound

(4.63)
$$\begin{aligned} v_T &\le c_5 q \gamma e^{-\beta' p} + c_6 \varepsilon^{-1/2} h_T^{-d/2} q \gamma e^{-\beta' p} \\ &\le c_7 \exp\left(-c_8 T^{\frac{r}{2r+d}-\zeta}\right) \end{aligned}$$

where c_7 and c_8 are strictly positive and $\zeta > 0$ arbitrarily small. Consequently the bound in (4.60) is asymptotically greater than the bound in (4.63), hence (4.54). ∎

The next lemma shows that (Z_T) satisfies condition (b) in Lemma 4.2.

LEMMA 4.4 (Z_T) *satifies the uniform Lipschitz condition*

(4.64) $$\sup_{x \in \mathbb{R}^d, \omega \in \Omega} |Z_T(x,\omega) - Z_S(x,\omega)| \le \Lambda |T - S|;$$

$T > 1$, $S > 1$; where Λ does not depend on (x, ω, S, T).

Proof
   We only prove (4.64) for

$$W_T = \frac{1}{\log_m T} \left(\frac{T}{\log T}\right)^{\frac{r}{2r+d}} f_T(x) \; ; \; T > 1$$

and with the constant $\dfrac{\Lambda}{2}$, since the result for EW_T is an easy consequence of this one, because

$$\begin{aligned} |EW_S - EW_T| &\le E|W_S - W_T| \\ &\le \sup_{\omega,x} |W_S - W_T|. \end{aligned}$$

Now we put

$$\log W_T = U_T + V_T$$

where $U_T = -\log_{m+1} T + \left(\dfrac{r+d}{2r+d}\right) \log \left(\dfrac{T}{\log T}\right) + \log \dfrac{1}{T}$

and $V_T = \log \left(\displaystyle\int_0^T K \left(\dfrac{x - X_t}{h_T}\right) dt\right)$ where the integral is supposed to be positive.

The derivative U_T' of U_T is clearly a $O\left(\dfrac{1}{T}\right)$.

Concerning V_T' first we have

$$\left(\int_0^T K \left(\frac{x - X_t}{h_T}\right) dt\right)' = \int_0^T \frac{\partial K}{\partial T} \left(\frac{x - X_t}{h_T}\right) dt + K \left(\frac{x - X_T}{h_T}\right).$$

Noting that

$$\frac{\partial K_0}{\partial T} \left(\frac{x_j - X_{t,j}}{h_T}\right) = -\frac{h_T'}{h_T^2}(x_j - X_{t,j}) K_0' \left(\frac{x_j - X_{t,j}}{h_T}\right) \; ; \; j = 1, \ldots, d$$

and that for some c_K, $K_0'(u) = 0$ if $|u| \geq c_K$ we obtain

(4.65) $$\left| \frac{\partial K_0}{\partial T} \left(\frac{x_j - X_{t,j}}{h_T}\right) \right| \leq \frac{|h_T'|}{h_T^2} a h_T \parallel K_0' \parallel_\infty \; ; \; j = 1, \ldots, d.$$

From (4.65) it is easy to deduce that

$$\left| \frac{\partial K}{\partial T} \left(\frac{x - X_t}{h_T}\right) \right| = O\left(\frac{1}{T}\right)$$

and finally

$$(\log W_T)' \leq \frac{c_1}{T} + \frac{c_2}{\int_0^T K \left(\frac{x - X_t}{h_T}\right) dt}.$$

Using the relation $W_T' = W_T (\log W_T)'$ it is then easy to find that

$$|W_T'| \leq \frac{c}{T^{\frac{r}{2r+d}}} \; , \quad T > 1$$

where c is constant. Thus W_T' is bounded hence (4.64). Clearly the result remains valid if $\int_0^T K \left(\frac{x - X_t}{h_T}\right) dt = 0$. ∎

We are now in a position to state a first consistency result :

THEOREM 4.7 *If $\alpha(u) \leq \gamma u^{-\beta}$, $\gamma > 0$, $\beta > \max \left(2\dfrac{p-1}{p-2}, \dfrac{7r+5d}{2r}\right)$ then*

(4.66) $$\frac{1}{\log_m T} \left(\frac{T}{\log T}\right)^{\frac{r}{2r+d}} |f_T(x) - f(x)| \underset{T \to \infty}{\longrightarrow} 0 \; a.s.,$$

$m \geq 1, x \in \mathbb{R}^d$

Proof

(4.53) implies

$$P(|Z_T(x)| > \eta) \;=\; O\!\left(\tfrac{1}{T^{1+\mu}}\right)$$

and (4.64) implies condition (b) in Lemma 4.2.
Hence (4.66) by using Lemma 4.2. ∎

We now state a uniform result

THEOREM 4.8 *If (X_t) is GSM then*

(4.67)
$$\sup_{\|x\| \le T^a} |Z_T(x)| \to 0 \quad a.s.$$

$m \ge 1, a > 0$.

Proof

Since K is clearly Lipschitzian we may use a method similar to the method of the proof in Theorem 2.2 : we take as $\|\cdot\|$ the sup norm and we construct a covering of $\{x : \| x \| \le T^a\}$ with ν_T^d hypercubes where $\nu_T \sim T^{a + \frac{r+d+1}{2r+d}}$. Thus we have

(4.68)
$$\sup_{\|x\| \le T^a} |Z_T(x)| \le \sup_{1 \le j \le \nu_T^d} |Z_T(x_{jT})| + O\left(\frac{1}{(\log T)^w}\right)$$

where the x_{jT}'s are the centers of the hypercubes and where $w > 0$. Using (4.54) we obtain

$$P\left(\sup_{1 \le j \le \nu_T^d} |Z_T(x_{jT})| > \eta\right) \le \frac{B\nu_T^d}{T^{C(\log_m T)^2}} = O\left(\frac{1}{T^{\log_m T}}\right).$$

On the other hand (4.64) shows that $T \mapsto \sup_{1 \le j \le \nu_T^d} |Z_T(x_{jT})|$ is uniformly continuous for each ω since Λ does not depend on (x, ω). Consequently we may apply Lemma 4.2 and we obtain (4.67) from (4.68). ∎

COROLLARY 4.5 *(Uniform optimal rate.)*
If f is ultimately decreasing with respect to $\|\cdot\|$, if (X_t) is GSM, if $\sup_{0 \le t \le T} \| X_t \|$ is measurable for each T and if $E\left(\sup_{0 \le t \le 1} \| X_t \|^a\right) < \infty$ for some $a > 0$ then

(4.69)
$$\frac{1}{\log_m T} \left(\frac{T}{\log T}\right)^{\frac{r}{2r+d}} \sup_{x \in \mathbb{R}^d} |f_T(x) - f(x)| \to 0 \quad a.s.$$

Proof

Since f is ultimately decreasing we have $\lim_{\|u\| \to \infty} \| u \| f(u) = 0$[1] hence it is

---

[1]To prove this it suffices to note that for R large enough

$$\int_{R/2 \le \|v\| \le R} f(v)dv \ge f(e_R)a_d R^d$$

where e_R denotes a vector such that $\|e_R\| = \dfrac{R}{2}$ and a_d is a positive constant.

easy to check that

$$(4.70) \qquad \frac{1}{\log_m T} \left(\frac{T}{\log T} \right)^{\frac{r}{2r+d}} \sup_{\|x\| > T^{a/2}} f(x) \to 0,$$

thus from Theorem 4.8 and (4.70) we deduce that it suffices to show that

$$\sup_{\|x\| > T^{a/2}} |Z_T(x)| \to 0 \text{ a.s.} .$$

To this aim we first note that $\sup_{0 \le t \le T} \| X_t \| \le \frac{T^{2a}}{2}$ and $\| x \| > T^{2a}$ imply
$\| \frac{x - X_t}{h_T} \| > \frac{T^{2a}}{2h_T}$, $0 \le t \le T$.
Now let c_K be such that $K(u) = 0$ if $\| u \| \ge c_K$ and let T_0 such that $\frac{T^{2a}}{2h_T} > c_K$ for, $T \ge T_0$.
We have $K\left(\frac{x - X_t}{h_T} \right) = 0$ for $T \ge T_0$, hence

$$\left\{ \sup_{0 \le t \le T} \| X_t \| \le \frac{T^{2a}}{2}, \| x \| > T^{2a} \right\} \Rightarrow \left\{ \sup_{\|x\| > T^{2a}} |Z_T(x)| = 0 \right\}.$$

Therefore for T large enough

$$P\left(\sup_{\|x\| > T^a} |Z_T(x)| > \eta \right) \le P\left(\sup_{0 \le t \le T} \| X_t \| > \frac{T^{2a}}{2} \right)$$
$$\le P\left(\sup_{0 \le t \le T} \| X_t \| > T^a \right), \ \eta > 0.$$

Now since $T \to \sup_{\|x\| > T^a} |Z_T(x)|$ is uniformly continuous for each ω, we may apply Lemma 4.2, hence (4.69). ∎

4.3.2 Superoptimal rate

We now state a result which shows that a full rate is also reached in the setting of uniform convergence.

We consider the hypothesis :

$$H : \| g_u \|_\infty \in L^1(]0, +\infty[)$$

then we have

THEOREM 4.9 *Under the conditions of Corollary 4.5 except that $A(\Gamma, \lambda)$ is replaced by H and that $h_T \sim T^{-\gamma}$ where $\frac{1}{2r} \le \gamma < \frac{1}{2d}$, we have for all $m \ge 1$*

$$(4.71) \qquad \frac{1}{\log_m T} \left(\frac{T}{\log T} \right)^{\frac{1}{2}} \sup_{x \in \mathbb{R}^d} |f_T(x) - f(x)| \to 0 \text{ a.s.}$$

Proof

As in the proof of Lemma 4.3 we consider decomposition (4.56) and we apply inequality (1.26).

The main task is to evaluate the variance of $\sum_{j=1}^{p} Y_{jn}$. First (4.39) yields

$$p\delta V\left(\frac{1}{p\delta}\int_0^{p\delta} K_{h_T}(x - X_t)dt\right)$$

$$\leq 2\int_{\mathbb{R}^d} |K_{h_T}(x - y)K_{h_T}(x - z)|\int_0^{+\infty} |g_u(y, z)|dudydz$$

$$\leq 2\int_0^{+\infty} \| g_u \|_\infty \, du \left(\int |K|\right)^2 =: M$$

thus

$$V\left(\sum_{j=1}^{p} Y_{jn}\right) \leq \frac{p}{\delta}M$$

therefore it is easy to see that

$$v^2(q) \leq \frac{2}{p\delta}M + \frac{\| K \|_\infty \, \varepsilon}{h_T^d}$$

then, choosing $p = \left[h_T^d\varepsilon^{-1}\right] + 1$ we obtain

$$v^2(q) \leq (2M + \| K \|_\infty)\varepsilon h_T^{-d}.$$

Consequently

$$\exp\left(-\frac{\varepsilon^2}{8v^2(q)}q\right) \leq \exp\left(-\frac{1}{8(2M + \| K \|_\infty)}\varepsilon h_T^d q\right).$$

We now choose $\varepsilon = \left(\frac{\log T}{T}\right)^{1/2}(\log_m T)^{1/2}\eta \quad (\eta > 0)$

and we note that $q = \frac{n}{2p} = \frac{T}{2p\delta}$ hence

$$\exp\left(-\frac{\varepsilon^2}{8v^2(q)}q\right) \leq \frac{1}{T^{c\log_m T}} \quad (c > 0)$$

where c depends only on M, $\| K \|_\infty$ and η.

Concerning the second term in the bound (1.26) it is easy to see that it takes the form

$$\exp\left(-c'\log\frac{1}{\rho}T^{\frac{1}{2}-\gamma d}\right) \quad (c' > 0).$$

Finally

$$P\left(\frac{1}{\log_m T}\left(\frac{T}{\log T}\right)^{\frac{1}{2}}|f_T(x) - Ef_T(x)| > \eta)\right) \le \psi_\eta(T)$$

where ψ_η is integrable on $]1, +\infty[$.

On the other hand it is easy to see that Lemma 4.4 remains valid if Z_T is replaced by

$$Z_{1,T} = \frac{1}{\log_m T}\left(\frac{T}{\log T}\right)^{1/2}(f_T(x) - Ef_T(x))$$

thus Lemma 4.2 implies that $Z_{1,T} \to 0$ a.s.

The bias is again given by (4.15) thus

$$\frac{1}{\log_m T}\left(\frac{T}{\log T}\right)^{1/2}|Ef_T(x) - f(x)| \le c_{(r)}\frac{T^{1/2-\gamma r}}{\log_m T \cdot (\log T)^{1/2}}$$

which tends to zero since $\gamma \ge \dfrac{1}{2r}$.

Finally uniform convergence is obtained by using the same process as in the proofs of Theorem 4.8 and Corollary 4.5. ∎

4.4   Sampling

In continuous time, data are often collected by using a sampling scheme. Various sampling designs can be employed. In the following we only consider three kinds of deterministic designs : dichotomy, irregular sampling, admissible sampling.

4.4.1   Dichotomy

Consider the data $(X_{jT/N} ; j = 1,\ldots, N)$ where $N = 2^n$; $n = 1, 2, \ldots$ T being fixed. Such a design may be associated with the accuracy of an instrument used for observing the process (X_t) over $[0, T]$.

In some parametric cases estimators based on that sampling are consistent. A well known example should be the observation of a Wiener process $(W_t, t \ge 0)$ at times jT/N. The associated estimator of the parameter σ^2 is

$$\sigma_N^2 = \frac{1}{T}\sum_{j=1}^{N}(W_{jT/N} - W_{(j-1)T/N})^2$$

which is clearly consistent in quadratic mean and almost surely.

Now if $(X_t, t \in \mathbb{R})$ is a process with identically distributed margins the density kernel estimator is

$$(4.72) \qquad \hat{f}_N(x) = \frac{1}{Nh_N^d} \sum_{j=1}^N K\left(\frac{x - X_{jT/N}}{h_N}\right) \ , \ x \in \mathbb{R}^d.$$

The following theorem shows that $\hat{f}_N$ is *not* consistent.

THEOREM 4.10 *let $(X_t, \ t \in \mathbb{R})$ be a zero mean real stationary Gaussian process with an autocorrelation function ρ satisfying*

$$0 < cu^\alpha \le 1 - \rho^2(u) \le c'u^\alpha \ , \ 0 < u \le T$$

where

$$0 < c \le c' < 1 \ and \ 0 < \alpha \le 2.$$

Then if $h_N = N^{-\gamma}$ $(0 < \gamma < 1)$ and if the kernel K satisfies $\int u^4 K(u) du < +\infty$ we have

$$(4.73) \qquad \liminf_{N \to +\infty} V\hat{f}_N(0) \ge \frac{4}{\pi\sqrt{c'}(2-\alpha)(4-\alpha)} - \frac{1}{2\pi} > 0.$$

In particular $V\hat{f}_N(0)$ tends to infinity if $\alpha = 2$.

Proof (sketch) :

We may and do suppose that $T = 1$ and $EX_0^2 = 1$. Now let us consider the decomposition

$$V\hat{f}_N(0) = \widetilde{V}_N + C_N + R_N + r_N$$

where $\widetilde{V}_N = \dfrac{1}{N^2 h_N^2} \sum_{j=1}^N VK\left(X_{j/N}/h_N\right)$,

$$C_N = -\frac{2}{N^2 h_N^2} \sum_{j=1}^{N-1} (N-j) \int K\left(\frac{u}{h_N}\right) K\left(\frac{v}{h_N}\right) f(u)f(v)dudv \ ,$$

$$R_N = \frac{2}{N} \sum_{j=1}^{N-1} \left(1 - \frac{j}{N}\right) f_{j/N}(0,0),$$

and

$$r_N = \frac{2}{N} \sum_{j=1}^{N-1} \int \left(1 - \frac{j}{N}\right) \left(f_{j/N}(h_N y, h_N z) - f_{j/N}(0,0)\right) K(y)K(z)dydz.$$

First, Bochner's lemma implies $\widetilde{V}_N \to 0$ and $C_N \to -f^2(0)$.

Now, since $f_{j/N}(0,0) = \dfrac{1}{2\pi}(1 - \rho^2(j/N))^{-\frac{1}{2}}$ we have

$$R_N \ge \frac{1}{\pi\sqrt{c'}} \frac{1}{N} \sum_{j=1}^{N-1} \left(1 - \frac{j}{N}\right) \left(\frac{N}{j}\right)^{\alpha/2}$$

which appears to be a Riemann sum for the function $\dfrac{1}{\pi\sqrt{c'}}(1-u)u^{-\alpha/2}$. Consequently

$$\liminf R_N \geq \frac{1}{\pi\sqrt{c'}}\left(\frac{1}{1-\dfrac{\alpha}{2}} - \frac{1}{2-\dfrac{\alpha}{2}}\right) \quad, 0 < \alpha \leq 2.$$

Finally by using $1 - \rho^2(j/N) \geq c\left(\dfrac{j}{N}\right)^{\alpha}$ and the inequality

$|e^{au} - 1| \leq au\left(1 + \dfrac{au}{2}\right)$ $(a > 0, u > 0)$ it is easy to check that r_N tends to zero.

Collecting the above results one obtains (4.73). ■

Under slightly different hypotheses it may be established that

$$(4.74) \qquad \lim_{N\to\infty} V\hat{f}_N(0) = \frac{1}{\pi T}\int_0^T \frac{1-u}{(1-\rho^2(u))^{1/2}}du.$$

In conclusion it appears that the condition $h_N \to 0$ is not appropriate in the dichotomy context. It is then necessary to adopt another point of view by considering $\hat{f}_N$ as an approximation of f_T and by letting h_N tend to h_T. Thus we have the following.

THEOREM 4.11 *If $(X_t, 0 \leq t \leq T)$ has cadlag sample paths, if K is uniformly continuous and if $h_N \to h_T$ then*

$$(4.75) \qquad \hat{f}_N(x) \xrightarrow[N\to\infty]{} f_T(x) \quad, x \in \mathbb{R}^d.$$

Proof

We have

$$\hat{f}_N(x) = \int_0^T K_{h_N}(x-u)d\mu_N(u)$$

and

$$f_T(x) = \int_0^T K_{h_T}(x-u)d\mu_T(u)$$

where $\mu_N = \dfrac{1}{N}\sum_{j=1}^N \delta_{(X_{jT/N})}$ and μ_T are empirical measures.

Now let φ be a continuous real function defined on $[0, T]$, then for all ω in Ω

$$\int_0^T \varphi d\mu_N = \frac{1}{T}\cdot\frac{T}{N}\sum_{j=1}^N \varphi(X_{jT/N}) \xrightarrow[N\to\infty]{} \frac{1}{T}\int_0^T \varphi(X_t)dt$$

since $t \mapsto \varphi \circ X_t(\omega)$ is Riemann integrable over $[0, T]$.
In particular

$$\int_0^T K_{h_T}(x-u)d\mu_N(u) \xrightarrow[N\to\infty]{} \int_0^T K_{h_T}(x-u)d\mu_T(u) = f_T(x).$$

On the other hand

$$\int_0^T \left(K_{h_T}(x - u) - K_{h_N}(x - u) \right) d\mu_N(u) \xrightarrow[N \to \infty]{} 0$$

since K is uniformly continuous. Hence (4.75). ∎

4.4.2 Irregular sampling

Consider the data $X_{t_1}, \ldots, X_{t_n}$ where $0 < t_1 < \ldots < t_n$ and $\min_{1 \le j \le n-1} (t_{j+1} - t_j) \ge m > 0$ for some m. The corresponding estimator is

$$(4.76) \qquad \overline{f}_n(x) = \frac{1}{nh_n^d} \sum_{j=1}^n K\left(\frac{x - X_{t_j}}{h_n} \right), \quad x \in \mathbb{R}^d.$$

Then it is not difficult to see that the asymptotic behaviour of $\overline{f}_n$ is the same as that of f_n studied in Chapter 2. Thus all the results in Chapter 2 remain valid with slight modifications.

4.4.3 Admissible sampling

We now consider a process $(X_t, t \in \mathbb{R})$ with irregular paths observed at sampling instants. In order to modelize the fact that the observations are frequent during a long time we assume that these sampling instants are $\delta_n, 2\delta_n, \ldots, n\delta_n$ where $\delta_n \to 0$ and $T_n = n\delta_n \to +\infty$.

Here the kernel estimator is defined as

$$(4.77) \qquad f_n^*(x) = \frac{1}{nh_n^d} \sum_{j=1}^n K\left(\frac{x - X_j \delta_n}{h_n} \right), \quad x \in \mathbb{R}^d.$$

Now we will say that (δ_n) is an *admissible sampling* if the superoptimal rate remains valid when the observations are $X_{\delta_n}, X_{2\delta_n}, \ldots, X_{n\delta_n}$ with a minimal sample size n.

More precisely (δ_n) is admissible if

(a) For a suitable choice of (h_n)

$$E(f_n^*(x) - f(x))^2 = O\left(\frac{1}{T_n} \right).$$

(b) δ_n is maximal (i.e. n is minimal) that is, if (δ_n') is a sequence satisfying (a) then $\delta_n' = O(\delta_n)$.

Note that if (δ_n) and (δ_n') are both admissible then obviously $\delta_n' \simeq \delta_n$.

In order to specify an admissible sampling we need the following assumptions

(1) $g_{s,t} = g_{|t-s|}$ exists for $s \neq t$ and $\| g_u \|_\infty \leq \pi(u)$, $u > 0$ where $(1 + u)\pi(u)$ is integrable over $]0, +\infty[$ and $u\pi(u)$ is bounded and ultimately decreasing. Furthermore $g_u(\cdot, \cdot)$ is continuous at (x, x).

(2)

$$\sup_{(y,z)\in\mathbb{R}^{2d}} \left| \int_0^{+\infty} g_u(y,z)du - \sum_{k=1}^{+\infty} \delta_n g_{k\delta_n}(y,z) \right| \xrightarrow[\delta_n \to 0]{} 0 \ .$$

These assumptions are satisfied if, for example, (X_t) is an ORNSTEIN-UHLENBECK process.

THEOREM 4.12 *If (1) and (2) hold, if $f \in C_r^d(\ell)$ and $K \in H_{k,\lambda}$ $(k + \lambda = r)$ then $\delta_n = T_n^{-d/2r}$ is admissible provided $h_n = T_n^{-1/2r}$.*

Proof
Let us begin with the following preliminary result :

(4.78)
$$\lim_{n\to\infty} \| H_n - G_n \|_\infty = 0$$

where $H_n(y,z) = \sum_{i=1}^\infty \delta_n g_{i\delta_n}(y,z)$

and $G_n(y,z) = \sum_{i=1}^{n-1} \left(1 - \frac{i}{n}\right) \delta_n g_{i\delta_n}(y,z)$.

In order to prove (4.78) note first that $u\pi(u)$ and $\pi(u)$ are decreasing for u large enough, $u > u_0$ say.
Therefore

$$\sum_{i-1 > \delta_n^{-1} u_0} |\delta_n g_{i\delta_n}(y,z)| \leq \sum_{i-1 > \delta_n^{-1} u_0} \delta_n \pi(i\delta_n)$$
$$\leq \int_{u_0}^{+\infty} \pi(u)du < +\infty.$$

On the other hand

$$|i\delta_n \pi(i\delta_n)| \leq \delta_n^{-1} \int_{(i-1)\delta_n}^{i\delta_n} u\pi(u)du \ , i - 1 > \delta_n^{-1} u_0.$$

Now we have

$$H_n - G_n = \sum_{i=n}^\infty \delta_n g_{i\delta_n} + \frac{1}{n} \sum_{i=1}^{n-1} i\delta_n g_{i\delta_n}$$

hence for n large enough

$$
\begin{aligned}
\| H_n - G_n \|_\infty &\le \sum_{i=n}^\infty \delta_n \pi(i\delta_n) + \frac{1}{n} \sum_{i=1}^{n-1} i\delta_n \pi(i\delta_n) \\
&\le \int_{(n-1)}^\infty \delta_n \pi(u)du + \frac{1}{n} \sum_{i-1\le \delta_n^{-1} u_0} i\delta_n \pi(i\delta_n) \\
&\quad + \frac{1}{n\delta_n} \int_{u_0}^{(n-1)\delta_n} u\pi(u)du \\
&\le \int_{(n-1)\delta_n}^{+\infty} \pi(u)du + \frac{u_0}{n\delta_n} \| u\pi(u) \|_\infty \\
&\quad + \frac{1}{n\delta_n} \int_{u_0}^{+\infty} u\pi(u)du,
\end{aligned}
$$

hence (4.78) since $\pi(u)$ and $(u\pi(u))$ are integrable and $n\delta_n \to \infty$.

We now study the variance of f_n by using the classical decomposition

$$
V f_n(x) = \widetilde{V}_n + C_n
$$

where $\widetilde{V}_n$ stands for the sum of variance and where

$$
C_n = \frac{2}{n\delta_n} \int K_{h_n}(x - y)K_{h_n}(x - z)G_n(y, z)dydz.
$$

For $\widetilde{V}_n$ we have again the well known result

(4.79)
$$
nh_n^d \widetilde{V}_n \to f(x) \int K^2.
$$

Concerning C_n note that

$$
\left| \frac{n\delta_n}{2}C_n - \int K_{h_n}(x - y)K_{h_n}(x - z)H_n(y, z)dydz \right| \le \| H_n - G_n \|_\infty
$$

and

$$
\left| \int K_{h_n}(x - y)K_{h_n}(x - z)[H_n(y, z) - G(y, z)]dydz \right| \le \| H_n - G \|_\infty
$$

where $G(y, z) = \int_0^{+\infty} g_u(y, z)du$.

Consequently assumption (2) and (4.78) entail

$$
\left| \frac{n\delta_n}{2}C_n - \int K_{h_n}(x - y)K_{h_n}(x - z)G(y, z)dydz \right| \to 0.
$$

Since G is continuous at (x, x) we find

(4.80)
$$n\delta_n C_n \to 2 \int_0^{+\infty} g_u(x, x) du.$$

Now the bias is given by (4.15) and (4.16), then by using (4.79) and (4.80) we obtain

$$E\left(f_n^*(x) - f(x)\right)^2 \le \frac{a}{nh_n^d} + a' h_n^{2r} + \frac{a''}{n\delta_n}$$

where a, a' and a'' are positive constants. Hence

$$E\left(f_n^*(x) - f(x)\right)^2 \le \frac{a'''}{n^{2r/(2r+d)}} + \frac{a''}{n\delta_n}$$

and since $n\delta_n = T_n$

$$E\left(f_n^*(x) - f(x)\right)^2 \le \frac{a''}{T_n} + a''' \left(\frac{\delta_n}{T_n}\right)^{\frac{2r}{2r+d}},$$

thus the full rate is obtained by choosing $\delta_n = T_n^{-d/2r}$ as announced above.

It remains to prove that δ_n is minimal : let us consider a sequence (δ_n') which generates the full rate and let us note that there exists $a_1 > 0$ such that

(4.81)
$$\frac{a_1}{T_n} \ge E\left(f_n^{**} - f(x)\right)^2 \ge \frac{a'''}{n^{2r/(2r+d)}} = a''' \left(\frac{\delta_n'}{T_n}\right)^{\frac{2r}{2r+d}}$$

where f_n^{**} is associated with the sampling $(X_{j\delta_n'})$.
Then (4.81) yields

$$\delta_n' \le \frac{a_1}{a'''} T_n^{-\frac{d}{2r}} = O(\delta_n)$$

and the proof of Theorem 4.12 is therefore complete. ∎

The following corollary provides the exact asymptotic quadratic error associated with an admissible sampling.

COROLLARY 4.6 *If (1) and (2) hold, if $f \in C_{2,1}(b)$ and if $f(x)f''(x) > 0$ then the choice $\delta_n = \lambda T_n^{-1/4}$ $(n > 0)$ and $h_n = \left(\frac{a\lambda}{b}\right)^{1/5} T_n^{-1/4}$ where $a = f(x) \int K^2$ and $b = f''^2(x) \left(\int u^2 K(u) du\right)^2$ leads to*

(4.82)
$$\lim_{n\to\infty} T_n E\left(f_n^*(x) - f(x)\right)^2 = 2 \int_0^{+\infty} g_u(x, x) du + \lambda^{4/5} \frac{5}{4} a^{4/5} b^{1/5}.$$

Proof :
Straightforward since the bias is given by (2.9). ∎

Note that if all the sample path $(X_t, 0 \leq t \leq T_n)$ is available (4.43) in Corollary 4.3 gives a smaller constant, namely

$$(4.83) \qquad \lim_{n \to \infty} T_n E \left(f_{T_n}(x) - f(x) \right)^2 = 2 \int_0^{+\infty} g_u(x, x) + \lambda^{4/5} \frac{1}{4} a^{4/5} b^{1/5}.$$

The reason is that a diagonal variance term appears in (4.80).

The following theorem shows that the superoptimal uniform convergence rate still remains valid if the sampling is admissible.

THEOREM 4.13 *Under the conditions of Theorems 4.8 and 4.9 where* $\delta_n \cong T_n^{-d/2r}$, $h_n^d = \delta_n$, *and* $r > d$:

$$(4.84) \qquad \frac{1}{\log_m T_n} \left(\frac{T_n}{\log T_n} \right)^{1/2} \sup_{x \in \mathbb{R}^d} |f_n^*(x) - f(x)| \xrightarrow[n \uparrow \infty]{} 0 \quad a.s.$$

Proof
  Let us consider the random variables

$$Z_{jn} = K_{h_n} \left(\frac{x - X_{j\delta_n}}{h_n} \right) - E K_{h_n} \left(\frac{x - X_{j\delta_n}}{h_n} \right), \quad 1 \leq j \leq n.$$

Then

$$f_n^*(x) - E f_n^*(x) = \frac{1}{n} \sum_{j=1}^{n} Z_{jn}.$$

As in Theorem 4.9 we use inequality (1.26).
First if $p\delta_n \to +\infty$ we may use the proof of Theorem 4.9. We obtain

$$V \left(\frac{1}{p} \sum_{j=1}^{p} Z_{jp} \right) \simeq \frac{1}{ph_n^d} + \frac{1}{p\delta_n}$$

consequently since $h_n^d = \delta_n$.

$$V \left(\sum_{j=1}^{p} Z_{jp} \right) \simeq \frac{p}{\delta_n}$$

hence since $|Z_{jn}| = O \left(h_n^{-d} \right) = O(\delta_n)$

$$v^2(q) \simeq \frac{1}{p\delta_n} + \frac{\varepsilon}{\delta_n}$$

and choosing $p \simeq \frac{1}{\varepsilon}$ we obtain $v^2(q) \simeq \frac{\varepsilon}{\delta_n}$.

Finally the first term of the bound in (1.26) has an exponent of the form $\dfrac{\varepsilon^2 q}{v^2(q)} \simeq$

$\dfrac{\varepsilon^2 q \delta_n}{\varepsilon}$ but $q = \dfrac{n}{2p} = \dfrac{T}{2p\delta_n} \sim \dfrac{T\varepsilon}{\delta_n}$, therefore $\dfrac{\varepsilon^2 q}{v^2(q)} \simeq T\varepsilon^2$.

Now choosing $\varepsilon = \left(\dfrac{T_n}{\log T_n}\right)^{1/2} \dfrac{1}{(\log_m T_n)}$ (a choice which is compatible with

$p\delta_n \to \infty$ since $\delta_n \sim T_n^{-d/2r}$ and $\dfrac{1}{2} > \dfrac{d}{2r} \Leftrightarrow r > d$), we get the bound

$$\exp\left(-c(\log T_n)(\log_m T_n)^2\right) = \dfrac{1}{T_n^{c \log_m T_n}}.$$

Now, since (X_t) is GSM, the second term is a

$$O\left(\exp\left(-c\log\dfrac{1}{\rho}p_n\right)\right) = O\left(\exp\left(-c\log\dfrac{1}{\rho}T_n^{\frac{1}{2}-\gamma}\right)\right)$$

where γ is arbitrarily small.

Finally

$$P\left(\dfrac{1}{\log_m T_n}\left(\dfrac{T_n}{\log T_n}\right)^{1/2}|f_n^*(x) - Ef_n^*(x)| > \eta\right)$$

$$= O\left(T_n^{-c' \log_m T_n}\right) = O\left(n^{-c'' \log n}\right)$$

since $n = \left[T_n^{\frac{2r+d}{2r}}\right]$. Hence the a.s. convergence.

Concerning the bias we have

$$\dfrac{1}{\log\log T_n}\left(\dfrac{T_n}{\log T_n}\right)^{1/2}|Ef_n^* - f| \leq \dfrac{T_n^{1/2}h_n^r}{(\log T_n)^{1/2}\log\log T_n}$$

$$= O\left(\dfrac{1}{(\log T_n)^{1/2}\log\log T_n}\right) \to 0$$

Finally in order to obtain an uniform result it suffices to use the same covering method as in Theorem 4.8 and to conclude as in Corollary 4.5. ∎

Notes

BANON (1978) was the first who considered density estimation in continuous time. In his pioneer work he studied the case of a stationary diffusion process by using a recursive estimator. Related results were obtained by BANON and NGUYEN (1978, 1981), NGUYEN (1979), NGUYEN and PHAM (1980, 1981).

Most of the above results are obtained under the so called Rosenblatt G_2-condition. The strong mixing case has been investigated by DELECROIX (1980).

Estimators based on δ-sequences are studied in PRAKASA RAO (1979, 1990). The special case of wavelets appears in LEBLANC (1995).

CASTELLANA and LEADBETTER (1986) have obtained the surprising $\frac{1}{T}$-rate (Theorem 4.4, 2).

Deterministic or random sampled data are considered by NGUYEN and PHAM (1981), MASRY (1983), PRAKASA RAO (1990).

Most of the results in this chapter seem to be new except of course Theorem 4.4, 2 and some simple results which belong to the folklore of density estimation in continuous time.

Chapter 5

Regression estimation and prediction in continuous time

Despite its great importance in practice, nonparametric regression estimation in continuous time has not been much studied up to now. The current chapter is perhaps the first general work on that topic.

The main results are similar to those obtained in the previous chapter about density estimation : if the process sample paths are irregular enough then a parametric rate appears in regression estimation. This fact remains valid for suitable sampled data and when nonparametric prediction is considered.

Optimal and superoptimal rates are studied in Sections 2, 3 and 5. Section 4 is devoted to limit in distribution. Section 6 deals with sampling and, finally, applications to forecasting appear in Section 7.

Several of the proofs are not detailed or are omitted since they are easy combinations of proofs given in Chapters 3 and 4.

5.1 The kernel regression estimator in continuous time

Let $Z_t = (X_t, Y_t)$, $(t \in \mathbb{R})$ be a $\mathbb{R}^d \times \mathbb{R}^{d'}$-valued measurable stochastic process defined on a probability space $(\Omega, \mathcal{A}, P)$. Let m be a Borelian function of $\mathbb{R}^{d'}$ into $\mathbb{R}$ such that $(\omega, t) \mapsto m^2(Y_t(\omega))$ is $P \otimes \lambda_T$-integrable for each positive T (λ_T stands for Lebesgue measure on $[0, T]$).

Assuming that the Z_t's have the same distribution with density $f_Z(x, y)$, we wish to estimate the regression function $E(m(Y_0) \mid X_0 = \bullet)$ given the data $(Z_t,\ 0 \le t \le T)$.

Consider the following functional parameters :

$$f(x) = \int_{\mathbb{R}^{d'}} f_Z(x,y)dy \quad , \quad x \in \mathbb{R}^d$$

and

$$\varphi(x) = \int_{\mathbb{R}^{d'}} m(y) f_Z(x,y)dy \quad , \quad x \in \mathbb{R}^d.$$

We may use f and φ for defining a version of the regression by setting

(5.1)
$$\begin{aligned} r(x) &= \varphi(x)/f(x) &\text{if}\quad f(x) &> 0 \\ &= Em(Y_0) &\text{if}\quad f(x) &= 0. \end{aligned}$$

Now let K be a d-dimensional convolution **kernel** (cf. Chapter 2), the **kernel regression estimator** is defined as

(5.2)
$$\begin{aligned} r_T(x) &= \varphi_T(x)/f_T(x) &\text{if}\quad f_T(x) &> 0 \\ &= \tfrac{1}{T}\int_0^T m(Y_t)dt &\text{if}\quad f_T(x) &= 0 \end{aligned}$$

where

(5.3)
$$f_T(x) = \frac{1}{T}\int_0^T K_{h_T}(x - X_t)dt$$

and

(5.4)
$$\varphi_T(x) = \frac{1}{T}\int_0^T m(Y_t)K_{h_T}(x - X_t)dt$$

with $\lim_{T \to \infty} h_T = 0(+)$.

Note that r_T may be written under the suggestive form

(5.5)
$$r_T(x) = \int_0^T p_{tT}(x)m(Y_t)dt$$

where

(5.6)
$$\begin{aligned} p_{tT}(x) &= K\left(\frac{x - X_t}{h_T}\right) \Big/ \int_0^T K\left(\frac{x - X_t}{h_T}\right)dt &\text{if}\quad f_T(x) &> 0 \\ &= \frac{1}{T} &\text{if}\quad f_T(x) &= 0. \end{aligned}$$

In the following, in order to simplify the exposition, we will suppose that K is a strictly positive kernel, unless otherwise stated.

5.2   Optimal asymptotic quadratic error

First we study the case where $m(Y_t)$ is supposed to be bounded. In fact we introduce a slightly more general assumption, namely.

B_0 - There exists a positive constant M such that

$$\sup_{0 \leq t \leq T < +\infty} E\left(m^2(Y_t) \mid \mathcal{B}_T\right) \leq M^2 \quad \text{a.s.}$$

where $\mathcal{B}_T = \sigma(X_t,\ 0 \leq t \leq T)$.

B_0 is clearly satisfied if $m(Y_t)$ is bounded but also in some special situations, for example if the processes $(m(Y_t))$ and (X_t) are independent. Another interesting case should be the model

(5.7) $$m(Y_t) = r(X_t) + \varepsilon_t \quad ;\quad t \in \mathbb{R}$$

where $r(X_t)$ is bounded, (ε_t) is a square integrable strictly stationary process and where (X_t) and (ε_t) are independent.

Now let us set

$$g^*_{s,t} = f_{(z_s, z_t)} - f_Z \otimes f_Z \ , \ s \neq t$$

and

$$G_{s,t}(x, x') = \int_{\mathbb{R}^{2d'}} m(y)m(y')g^*_{s,t}(x, y\,;\,x', y')\, dy\, dy',$$

$(x, x') \in \mathbb{R}^{2d}$.

Furthermore we put

$$\Delta_{p^*}(\Gamma^*) = \sup_{(s,t) \notin \Gamma^*} \| G_{s,t} \|_{L^{p^*}(dx\, dx')}$$

where Γ^* is some two dimensional Borel set and $2 < p^* \leq +\infty$.

On the other hand $\alpha^{(2)}_*$ will denote the $2 - \alpha$-mixing coefficient of (Z_t) that is

$$\alpha^{(2)}_*(u) = \sup_{t \in \mathbb{R}} \alpha(\sigma(Z_t),\ \sigma(Z_{t+u})) \ ,\ u \geq 0.$$

With the above notations we may introduce the following conditions :

$A^*(\Gamma^*, p^*)$. There exist $\Gamma^* \in \mathcal{B}_{\mathbb{R}^2}$, containing $D = \{(s,t) \in \mathbb{R}^2 : s = t\}$, and $p^* \in]2, +\infty]$ such that

a) $g^*_{s,t}$ and $G_{s,t}$ exist for $(s,t) \notin \Gamma^*$

b) $\Delta_{p^*}(\Gamma^*) < +\infty$

c) $\limsup_{T \to \infty} \dfrac{1}{T} \displaystyle\int_{[0,T]^2 \cap \Gamma^*} ds\, dt = \ell_{\Gamma^*} < \infty.$

$\underline{M^*(\gamma, \beta)}$. $\alpha_*^{(2)}(|t - s|) \leq \gamma |t - s|^{-\beta}$; $(s, t) \notin \Gamma^*$ where $\gamma > 0$, $\beta > 0$.

$\underline{C^*}$. $f_Z \in C_{2,d+d'}(b)$, $f \in C_{2,d}(b')$, $\varphi \in C_{2,d}(b'')$ for some b, b', b''.

Note that, if m is bounded, these conditions may be simplified. In particular if $f_Z \in C_{2,d+d'}(b)$ then $f \in C_{2,d}(b)$ and $\varphi \in C_{2,d}(b \parallel m \parallel_\infty)$. Note in addition that we will also use some assumptions introduced in Chapter 4.

We are now in a position to state the "optimal rate" result

THEOREM 5.1 *If B_0, $A(\Gamma, p)$, $A^*(\Gamma^*, p^*)$, C^* and $M^*(\gamma, \beta)$ hold with*

$$\beta > \frac{\min(p, p^*) - 1}{\min(p, p^*) - 2} \text{ then the choice } h_T = c\,T^{-1/(d+4)}, \; c > 0, \text{ entails}$$

$$(5.8) \qquad \qquad \limsup_{T \to \infty} T^{4/(d+4)} \, E(r_T(x) - r(x))^2 \leq c(x)$$

provided $f(x) > 0$. $c(x)$ is an explicit constant.

If $p = p^* = +\infty$ the condition for β becomes $\beta > 2$ and the rate $T^{-4/(d+4)}$ remains valid for $\beta = 2$.

Proof
We first derive a preliminary result, namely

$$(5.9) \qquad \qquad E\left(r_T^2(f_T - Ef_T)^2\right) \leq M^2 V f_T$$

where x is omitted.

For that purpose by using (5.5), the Cauchy-Schwarz inequality and B_0 we obtain

$$E\left[r_T^2(f_T - Ef_T)^2\right] = E\left[E\left(r_T^2(f_T - Ef_T)^2 \mid \mathcal{B}_T\right)\right]$$

$$= E\left[(f_T - Ef_T)^2 E\left[\left(\int_0^T p_{tT} m(Y_t) dt\right)^2 \mid \mathcal{B}_T\right]\right]$$

$$= E\left[\left((f_T - Ef_T)^2 \int_{[0,T]^2} p_{sT} p_{tT} E(m(Y_s) m(Y_t) \mid \mathcal{B}_T)\right) ds\, dt\right]$$

$$\leq E\left[(f_T - Ef_T)^2 \left(\int_0^T p_{tT}\left[E(m^2(Y_t) \mid \mathcal{B}_T)\right]^{1/2} dt\right)^2\right]$$

$$\leq M^2 E(f_T - Ef_T)^2$$

which proves (5.9).

Now from the decomposition

$$r_T - \frac{E\varphi_T}{Ef_T} = r_T \frac{Ef_T - f_T}{Ef_T} + \frac{\varphi_T - E\varphi_T}{Ef_T}$$

and (5.9) we get

$$(5.10) \qquad \qquad E\left(r_T - \frac{E\varphi_T}{Ef_T}\right)^2 \leq \frac{2M^2}{(Ef_T)^2} V f_T + \frac{2}{(Ef_T)^2} V \varphi_T.$$

By Theorem 4.2 we have

$$(5.11) \qquad \limsup_{T\to\infty} T h_T^d V f_T(x) \leq \ell_\Gamma f(x) \int K^2.$$

Concerning $V\varphi_T$ we may use the same method as for $V f_T$ in Theorem 4.2, we obtain

$$(5.12) \qquad \limsup_{T\to\infty} T h_T^d V\varphi_T(x) \leq M^2 \ell_{\Gamma^*} f(x) \int K^2.$$

It remains to study the "pseudo-bias"

$$r - \frac{E\varphi_T}{E f_T} = r\,\frac{E f_T - f}{E f_T} + \frac{\varphi - E\varphi_T}{E f_T}.$$

Using C^* and classical methods one easily obtains

$$(5.13) \qquad \limsup_{T\to\infty} h_T^4 \left(r - \frac{E\varphi_T}{E f_T} \right)^2 \leq \frac{2r^2}{f^2}\,\frac{\chi_f^2}{4} + \frac{2}{f^2}\,\frac{\chi_\varphi^2}{4}$$

where

$$\chi_f^2 = \left(\sum_{1\leq i,j\leq d} \frac{\partial^2 f}{\partial x_i \partial x_j} \int u_i u_j K(u)du \right)^2$$

and χ_φ^2 is similar.

Finally collecting (5.10), (5.11), (5.12) and (5.13) we get (5.8) with

$$(5.14) \qquad \begin{aligned} c(x) &= \frac{2(1+M^2)}{c^d f(x)}(\ell_\Gamma + \ell_{\Gamma^*}) \int K^2 + \frac{c^4}{2f^2(x)}(r^2(x)\chi_f^2(x) \\ &+ \chi_\varphi^2(x)). \blacksquare \end{aligned}$$

We now turn to a more general case, replacing B_0 by

B_0' - $E(\exp a|m(Y_t)|) \leq M'$ for some $a > 0$ and some $M' > 0$.

First, it can be established that (5.8) remains valid, provided

$\beta > \max\left(2\,\dfrac{\min(p^*,p) - 1}{\min(p^*,p) - 2}, 2 + d \right)$; the proof uses arguments similar to those in Theorems 3.1 and 4.2.

Now we introduce the family $\mathcal{Z}$ of processes $Z = (Z_t, t \in \mathbb{R})$ which satisfy the above hypotheses uniformly with respect to a, M', Γ, p, Γ^*, p^*, γ, β, b, b', b'' and we consider a kernel of the type $K = K_0^{\otimes d}$, then we have

COROLLARY 5.1

$$(5.15) \qquad \lim_{T\to\infty} \sup_{Z\in\mathcal{Z}} T^{4/(d+4)} E_Z(r_T(x) - r_{(Z)}(x))^2 = c_{\mathcal{Z}}$$

where $c_{\mathcal{Z}}$ is explicit and $r_{(Z)}$ denotes the regression associated with Z.

The proof of Corollary 5.1 is analogous to the proofs of Corollaries 5.1 and 5.2. In particular, given a sequence (U_n, V_n), $n \in \mathbb{Z}$ of i.i.d. $\mathbb{R}^{d+d'}$-valued random variables one may construct the process

$$Z_t = (X_t, Y_t) = \left(U_{[t/L_0]}, V_{[t/L_0]} \right) , \quad t \in \mathbb{R} ,$$

then for a suitable choice of L_0 and (U_n, V_n), $n \in \mathbb{Z}$, $(Z_t, \ t \in \mathbb{R})$ belongs to Z and satisfies

(5.16) $$E_Z \big(r_T(x) - r_{(Z)}(x) \big)^2 \sim \frac{c_Z}{T^{4/(d+4)}} ,$$

details are omitted.

(5.16) shows that the "optimal rate" is achieved and (5.15) that better rates are feasible. These rates are considered in the next section.

5.3   Superoptimal asymptotic quadratic error

We now show that, if the sample paths are irregular enough, the kernel estimator exhibits a parametric asymptotic behaviour.

In order to express that irregularity we need some notations. Consider

$$g_{s,t} = f_{(X_s, X_t)} - f \otimes f \ ; \ s \neq t$$

and suppose that $g_{s,t} = g_{|t-s|}$, then we put

$$h(x', x'') = \int_{]0, +\infty[} |g_u(x', x'')| du .$$

Similarly if $g^*_{s,t} = g^*_{|s-t|}$, $s \neq t$, we put

$$H(x', x'') = \int_{]0, +\infty[} |G_u(x', x'')| du$$

where

$$G_u(x', x'') = \int_{\mathbb{R}^{2d'}} m(y) m(y') g^*_u(x, y; x', y') \, dy \, dy'.$$

Now the "irregularity" assumption is

I_1 - h and H exist and are continuous at (x, x).

The following theorem gives the parametric rate

THEOREM 5.2 *If B_0 and I_1 hold and if $f(x)$ is strictly positive then*

(5.17) $$\limsup_{T \to \infty} T E \left(r_T(x) - \frac{E \varphi_T(x)}{E f_T(x)} \right)^2 \leq c_1(x) ,$$

if in addition C^* holds then the choice $h_T = c T^{-1/4}$, $c > 0$ entails

(5.18) $$\limsup_{T \to \infty} T E (r_T(x) - r(x))^2 \leq c_2(x) .$$

$c_1(x)$ and $c_2(x)$ are explicit.

Proof

We first study $V\varphi_T(x)$. According to (5.4) we have

$$(5.19) \qquad TV\varphi_T(x) = \frac{2}{h_T^{2d}} \int_0^T \left(1 - \frac{u}{T}\right) Cov\left(m(Y_0)K\left(\frac{x - X_0}{h_T}\right),\right.$$
$$\left. m(Y_u)K\left(\frac{x - X_u}{h_T}\right)\right) du,$$

where the covariance, say γ_u, may be written

$$\gamma_u = \int m(y_1)\,m(y_2)\,K\left(\frac{x - x_1}{h_T}\right)K\left(\frac{x - x_2}{h_T}\right) g_u^*(x_1, y_1; x_2, y_2) dx_1\, dx_2\, dy_1\, dy_2$$
$$= \int_{\mathbb{R}^{2d}} K\left(\frac{x - x_1}{h_T}\right)K\left(\frac{x - x_2}{h_T}\right) G_u(x_1, x_2)\, dx_1\, dx_2,$$

therefore

$$TV\varphi_T(x) \le 2 \int_{\mathbb{R}^{2d}} K_{h_T}(x - x_1)\, K_{h_T}(x - x_2) \int_0^{+\infty} |G_u(x_1, x_2)|\, du$$

using I_1 we obtain

$$(5.20) \qquad \limsup_{T\to\infty} TV\varphi_T(x) \le 2 \int_0^{+\infty} |G_u(x, x)| du.$$

Now, by I_1, (4.35) is valid, thus

$$\limsup_{T\to\infty} TVf_T(x) \le 2 \int_0^{+\infty} |g_u(x, x)| du$$

and (5.10) implies

$$(5.21) \qquad \begin{aligned} \limsup_{T\to\infty} & TE\left(r_T(x) - \frac{E\varphi_T(x)}{Ef_T(x)}\right)^2 \\ & \le \frac{4(M^2 + 1)}{f(x)} \int_0^{+\infty} (|G_u(x, x)| + |g_u(x, x)|) du \end{aligned}$$

hence (5.17).

Concerning (5.18) it is an easy consequence of (5.13) and of the choice $h_T = cT^{-1/4}$. ∎

Under stronger conditions it is possible to substitute "lim" for "lim sup" in (5.18). To this aim let us define the function

$$g_{s,t}^{**}(x', x'', y) = f_{(X_0, Z_u)}(x', x'', y) - f(x')f_Z(x'', y)$$

and supposing that $g_{s,t}^{**} = g_{|t-s|}^{**}$ let us set

$$J_u(x', x'') = \int_{\mathbb{R}^{d'}} m(y)g_u^{**}(x', x'', y)\, dy \quad, \quad u > 0.$$

Now we need the following assumption :

I_2 - g_u, G_u, J_u exist, are bounded, continuous at (x,x), and $\| g_u \|_\infty$, $\| G_u \|_\infty$ and $\| J_u \|_\infty$ are integrable over $]0, +\infty[$.

We then have the following result

THEOREM 5.3 *If $m(Y_0)$ is bounded, if C^* and I_2 hold and if (Z_t) is GSM then $f(x) > 0$ and the choice $h_T = c\,T^{-1/4}$, $c > 0$ leads to*

$$(5.22) \qquad\qquad T \cdot E(r_T(x) - r(x))^2 \longrightarrow C^2(x)$$

where

$$
\begin{aligned}
(5.23) \quad C^2(x) \;=\; & c^4 \left(\sum_{1 \le i,j \le d} \left(\frac{\partial^2 r}{\partial x_i \partial x_j} + 2\,\frac{\partial \mathrm{Log} f}{\partial x_i}\,\frac{\partial r}{\partial x_j} \right)(x) \int u_i\,u_j\,K(u)\,du \right)^2 \\
& + \frac{2}{f(x)^2} \left(r^2(x) \int_0^{+\infty} g_u(x,x)\,du + \int_0^{+\infty} G_u(x,x)\,du \right. \\
& \left. - \varphi(x) \int_0^{+\infty} J_u(x,x)\,du \right).
\end{aligned}
$$

The proof is a combination of the proofs of Theorems 3.1 and 4.4.2 and is therefore omitted.

Now, to complete this section we state a result which offers intermediate rates. The main assumption is

$A''(p)$ - $G(s,t)$ exists for $s \ne t$, $\| G_{s,t} \|_p$ is locally integrable and

$$\limsup_{T \to \infty} \frac{1}{T} \int_{[0,T]^2} \| G_{s,t} \|_p \, ds\, dt = G_p < +\infty.$$

In $A''(p)$, p belongs to $[1, +\infty]$. In the case where $G_{s,t} = G_{|t-s|}$, $A''(p)$ is satisfied as soon as $\| G_u \|_p$ is integrable. In particular I_2 implies $A''(+\infty)$.

Intermediate rates depend on p and are specified in the following statement :

THEOREM 5.4 *Under the conditions B_0, $A'(p)$, $A''(p)$ and C^*, and if $h_T = cT^{-p/(4p+2d)}$ and $f(x) > 0$ then*

$$(5.24) \qquad\qquad \limsup_{T \to \infty} T^{2p/(2p+d)}\,E(r_T(x) - r(x))^2 \le D(x)$$

where $D(x)$ is explicit.

Proof

Owing to Theorem 4.6 and formulas (5.10) and (5.13) we only have to study

$$V\varphi_T(x) = \frac{1}{T^2} \int_{[0,T]^2 \times \mathbb{R}^{2d}} K_{h_T}(x - x_1)\,K_{h_T}(x - x_2)\,G_{s,t}(x_1, x_2)\,ds\,dt\,dx_1\,dx_2.$$

Supposing that $1 < p < \infty$ and using Hölder inequality, we arrive at

$$V\varphi_T(x) \leq \frac{1}{T^2} \left(\int_{\mathbb{R}^d} K_{h_T}^q (x - x') dx' \right)^{2/q} \int_{[0,T]^2} \| G_{s,t} \|_p \, ds \, dt$$

where $q = \dfrac{p}{p-1}$, hence

$$Th_T^{(2d)/p} V\varphi_T(x) \leq \frac{1}{T} \int_{[0,T]^2} \| G_{s,t} \|_p \, ds \, dt \cdot \| K \|_q^2$$

and taking the lim sup on both sides we get

(5.25) $$\limsup_{T \to \infty} Th_T^{(2d)/p} V\varphi_T(x) \leq G_p \| K \|_q^2$$

and the rest is clear. The special cases $p = 1$ and $p = \infty$ may be treated similarly. ∎

Note that the optimal rate is reached for $p = 2$ while the superoptimal rate is achieved for $p = +\infty$.

5.4 Limit in distribution

In order to specify the asymptotic distribution of r_T we introduce some notations (where x is omitted)

$$u = \frac{1}{f} \begin{bmatrix} 1 \\ -r \end{bmatrix}$$

$$A_T = \begin{bmatrix} V\varphi_T & Cov(f_T, \varphi_T) \\ Cov(f_T, \varphi_T) & V f_T \end{bmatrix}$$

moreover we suppose that $d = 1$ and that $Th_T A_T \to L$ a constant regular matrix.

On the other hand we set

$$\Pi_T = [v \, w] A_T \begin{bmatrix} v \\ w \end{bmatrix}$$ where v and w are real numbers.

Then we have the following weak convergence result :

THEOREM 5.5 *If C^* hold, $f(x) > 0$ and $\alpha(u) = O(e^{-\gamma u})$ $(\gamma > 0)$ then the choice $h_T = cT^{-\lambda} \left(c > 0, \frac{1}{5} < \lambda < \frac{1}{3} \right)$ entails*

(5.26) $$\frac{r_T(x) - r(x)}{\sqrt{(u' A_T u)(x)}} \xrightarrow{w} N$$

where N has a standard normal distribution.

(5.26) is an extension of a Schuster's result obtained in the i.i.d. case. Proof is omitted.

A confidence interval can be constructed from the following corollary :

COROLLARY 5.2

(5.27) $$\sqrt{Th_T} \sqrt{\frac{f_T(x)}{V_T(x)}} \, (r_T(x) - r(x)) \xrightarrow{w} N$$

where

(5.28) $$V_T(x) = \frac{1}{f_T(x)} \, \frac{1}{Th_T} \int_0^T m^2(Y_t) \, K\left(\frac{x - X_t}{h_T}\right) dt - r_T^2(x).$$

It should be noticed that **asymptotic normality of** f_T may be obtained from (5.26) or (5.27) (see [CP]).

5.5   Uniform convergence rates

We will now discuss uniform convergence. For this purpose we use a kernel $K = K_0^{\otimes d}$ where K_0 has compact support and continuous derivative. Then, if the functional parameters are twice continuously differentiable the obtained rates appear to be the same as in the density case as soon as the sup norm is taken on a compact set, say Δ, such that $\inf_{x \in \Delta} f(x) > 0$.

We summarize the results about optimal and superoptimal rates in the following theorem :

THEOREM 5.6 *Suppose that $m(Y_0)$ is bounded, C^* hold and (Z_t) is GSM then*

  1) If $A(\Gamma, p)$ and $A^*(\Gamma^*, p^*)$ hold, the choice $h_T \simeq T^{-1/(4+d)}$ entails for each $k \geq 1$

(5.29) $$\frac{1}{\text{Log}_k T} \left(\frac{T}{\text{Log} T}\right)^{2/(4+d)} \sup_{x \in \Delta} |r_T(x) - r(x)| \longrightarrow 0 \quad a.s..$$

  2) If I_1 hold, if $d = 1$, and if $h_T \simeq T^{-\gamma}$ where $\frac{1}{4} \leq \gamma < \frac{1}{2}$ then for each $k \geq 1$

(5.30) $$\frac{1}{\text{Log}_k T} \left(\frac{T}{\text{Log} T}\right)^{1/2} \sup_{x \in \Delta} |r_T(x) - r(x)| \longrightarrow 0 \quad a.s..$$

Proof (sketch)

Let us consider the decomposition

$$r_T - \frac{E\varphi_T}{Ef_T} = r_T \frac{Ef_T - f_T}{Ef_T} + \frac{\varphi_T - E\varphi_T}{Ef_T}$$

and let us set $M = max(1, \| \, m(Y_0) \, \|_\infty)$ and $\eta = \inf_{x \in \Delta} f(x)$, then for T large enough we have

(5.31)
$$\sup_{x \in \Delta} \left| r_T(x) - \frac{E\varphi_T(x)}{E f_T(x)} \right|$$

$$\leq \frac{2M}{\eta} \left(\sup_{x \in \Delta} |f_T(x) - Ef_T(x)| + \sup_{x \in \Delta} |\varphi_T(x) - E\varphi_T(x)| \right) \, .$$

Now, under the conditions in 1), Theorem 4.9 implies

$$\frac{1}{\text{Log}_k T} \left(\frac{T}{\text{Log} T} \right)^{2/(4+d)} \sup_{x \in \Delta} |f_T(x) - E f_T(x)| \longrightarrow 0 \text{ a.s..}$$

A similar result may be established for φ_T. This can be done by using the same scheme as in the density case (cf. Lemma 4.2, Lemma 4.3 and Lemma 4.4). One finally obtains

$$\frac{1}{\text{Log}_k T} \left(\frac{T}{\text{Log} T} \right)^{2/(4+d)} \sup_{x \in \Delta} \left| r_T(x) - \frac{E\varphi_T(x)}{E f_T(x)} \right| \longrightarrow 0 \text{ a.s.}$$

and (5.29) follows from C^*.

The proof of (5.30) is similar. ∎

5.6   Sampling

This section will be short because the reader can easily guess that regression and density estimators behave alike when sampled data are available. Consequently the results in section 4.4 remain valid.

In particular if data are constructed by dichotomy, that is by considering $X_{T/2^n}$, $X_{2T/2^n}$, ... , X_T the kernel regression estimator is not consistent under natural assumptions.

If the data are $X_{t_1}, \ldots, X_{t_n}$ with $0 < t_1 < \ldots < t_n$ and $\min_{1 \leq j \leq n-1} (t_{j+1} - t_j) \geq m > 0$ then the asymptotic quadratic and uniform errors are the same as that of r_n studied in Chapter 3.

We now consider a process $(Z_t, t \in \mathbb{R})$, with irregular paths, observed at times $\delta_n, 2\delta_n, \ldots, n\delta_n$ where $\delta_n \to 0$ and $T_n = n\delta_n \to \infty$. The associated kernel estimator is

$$(5.32) \qquad \widehat{r}_n(x) = \frac{\displaystyle\sum_{j=1}^n m\left(Y_{j\delta_n}\right) K\left(\frac{x - X_{j\delta_n}}{h_n} \right)}{\displaystyle\sum_{j=1}^n K\left(\frac{x - X_{j\delta_n}}{h_n} \right)} , \quad x \in \mathbb{R}^d.$$

In the same way as in subsection 4.4.3 we will say that (δ_n) is an **admissible sampling** if

(a) for a suitable choice of (h_n)

$$E(r_n(x) - r(x))^2 = O\left(\frac{1}{T_n} \right).$$

(b) δ_n is maximal (i.e. n is minimal) that is, if (δ'_n) is a sequence satisfying (a) then $\delta'_n = O(\delta_n)$.

Then under conditions similar to these of Theorems 4.13 and 5.6 it may be proved that $\delta_n = T_n^{-d/4}$ is admissible provided $h_n \simeq T_n^{-1/4}$, and that

$$\frac{1}{\mathrm{Log}_k T_n} \left(\frac{T_n}{\mathrm{Log} T_n} \right)^{1/2} \sup_{x \in \Delta} |\widehat{r}_n(x) - r(x)| \longrightarrow 0 \ \text{a.s.}$$

where Δ is any compact set such that $\inf_{x \in \Delta} f(x) > 0$.

5.7 nonparametric prediction in continuous time

Let $(\xi_t, \ t \in \mathbb{R})$ be a strictly stationary measurable process. Given the data $(\xi_t, \ 0 \leq t \leq T)$ we would like to predict the non-observed square integrable real random variable $\zeta_{T+H} = m(\xi_{T+H})$ where the horizon H satisfies $0 < H < T$ and where m is measurable and bounded on compact sets.

In order to simplify the exposition we suppose that (ξ_t) is a **real Markov process** with sample paths which are **continuous on the left**.

Now let us consider the associated process

$$Z_t = (X_t, Y_t) = (\xi_t, m(\xi_{t+H})), \ t \in \mathbb{R}$$

and the kernel regression estimator based on the data $(Z_t, \ 0 \leq t \leq T - H)$. The nonparametric predictor is

$$\widehat{\zeta}_{T+H} = \varphi_{T-H}(\xi_T)/f_T(\xi_T)$$

that is

$$(5.33) \qquad \widehat{\zeta}_{T+H} = \frac{\int_0^{T-H} m(\xi_{t+H}) K \left(\frac{\xi_T - \xi_T}{h_T} \right) dt}{\int_0^T K \left(\frac{\xi_T - \xi_t}{h_T} \right) dt} = \widehat{r}_T(\xi_T) \ ,$$

where the kernel K has a compact support S_K, is strictly positive over $\overset{\circ}{S}_K$ and has continuous derivative. Note that these conditions together with left continuity of paths entails that the denominator in (5.33) is strictly positive with probability 1.

We now study the asymptotic behaviour of $\widehat{\zeta}_{T+H}$ as T tends to infinity, H remaining fixed. As usual $\widehat{\zeta}_{T+H}$ is an approximation of $r(\xi_T) = E(\zeta_{T+H} \mid \xi_s, s \leq T) = E(\zeta_{T+H} \mid \xi_T)$.

If the sample paths of (ξ_t) are regular, the rates are similar to those obtained in Chapter 3, specifically in Theorem 3.5 and Corollary 3.1. We therefore focus our attention on the superoptimal case in order to exhibit sharper rates.

Let us first indicate the almost sure convergence rate.

COROLLARY 5.3 *If I_2 and C^* hold, (ξ_t) is GSM and if one chooses $h_T \simeq T^{-\gamma}$ where $\dfrac{1}{4} \leq \gamma < \dfrac{1}{2}$ then*

$$(5.34) \qquad \frac{1}{\mathrm{Log}_k T} \left(\frac{T}{\mathrm{Log}T} \right)^{1/2} [\widehat{r}_T(\xi_T) - r(\xi_T)] \, \mathbf{I}_{\xi_T \in \Delta} \xrightarrow{a.s.} 0$$

for each integer $k \geq 1$ and each compact set Δ such that $\inf_{x \in \Delta} f(x) > 0$.

Proof (sketch)
We have

$$(5.35) \qquad |\widehat{r}_T(\xi_T) - r(\xi_T)| \mathbf{I}_{\xi_T \in \Delta} \leq \sup_{x \in \Delta} \left| \frac{\varphi_{T-H}(x)}{f_T(x)} - r(x) \right| \quad a.s.$$

hence (5.34) using the same method as in Theorem 5.6. ∎

We presently turn to convergence in mean square. First we have the following results :

COROLLARY 5.4 *If conditions in Corollary 5.3 hold and if $h_T \simeq T^{-1/8}$ then*

$$(5.36) \qquad E\left[(r_{T+H}(\xi_T) - r(\xi_T))^2 \mathbf{I}_{\xi_T \in \Delta} \right] = 0(T^{-1/2})$$

for each closed interval Δ such that $\inf_{x \in \Delta} f(x) > 0$.

Proof
Using (5.13), (5.31) and (5.35) it is easy to realize that it is enough to study the asymptotic behaviour of $\delta_T = E\left(\sup_{x \in \Delta} |f_T(x) - Ef_T(x)|^2 \right)$ and $\delta'_T = E\left(\sup_{x \in \Delta} |\varphi_T(x) - E\varphi_T(x)|^2 \right)$. We only consider δ_T since δ'_T can be treated similarly.

Now we may and do suppose that $\Delta = [0,1]$, then using the condition $|K(x'') - K(x')| \leq \ell |x'' - x'|$ where $\ell = \| K'(\cdot) \|_\infty$ we obtain

$$(5.37) \qquad \sup_{x \in \Delta} |f_T(x) - Ef_T(x)| \leq \sup_{1 \leq j \leq k_T} |f_T(x_j) - Ef_T(x_j)| + \frac{2\ell}{k_T h_T^2} \, ,$$

where $x_j = \dfrac{j}{k_T}$, $1 \leq j \leq k_T$ and $k_T = [T^{1/2}]$.

Now (5.37) implies

$$(5.38) \qquad E\delta_T^2 \leq 2E\left(\sup_{1 \leq j \leq k_T} |f_T(x_j) - Ef_T(x_j)|^2 \right) + \frac{8\ell^2}{k_T^2 h_T^4}$$

which in turn implies

$$E\delta_T^2 \leq 2 \sum_{j=1}^{k_T} E|f_T(x_j) - Ef_T(x_j)|^2 + \frac{8\ell^2}{k_T^2 h_T^4}.$$

From I_2 and (4.39) we infer that

$$E\delta_T^2 \le 2\frac{k_T}{T} \int_0^{+\infty} \| g_u \|_\infty \, du + \frac{8\ell^2}{k_T^2 h_T^4}$$

thus

$$E\delta_T^2 = O(T^{-1/2})$$

and since the bias term is a $O(T^{-1/2})$ too, (5.36) follows. ∎

The last result requires a stronger assumption : let us suppose that (ξ_t) is φ_{rev}-mixing (cf. subsection 3.3.3) and consider the predictor defined for T large enough by

$$\zeta_{T+H}^* = r_{T'}(\xi_T)$$

where $T' = T - H - \mathrm{Log}T \cdot \mathrm{Log}_2 T$.

Then we have the following superoptimal rate :

COROLLARY 5.5 *If I_2 and C^* hold, if (ξ_t) is φ_{rev}-mixing with $\varphi_{rev}(p) \le a\rho^p$ $(a > 0, \ 0 < \rho < 1)$ then the choice $h_T \simeq T^{-1/4}$ entails*

$$(5.39) \qquad E\left[(r_{T'}(\xi_T) - r(\xi_T))^2 \mathbf{I}_{\xi_T \in \Delta}\right] = O\left(\frac{1}{T}\right)$$

for each compact set Δ.

Proof

First we have

$$\begin{aligned}
D_T &= \ : E\left[(r_{T'}(\xi_T) - r(\xi_T))^2 \mathbf{I}_{\xi_T \in \Delta}\right] \\
&= \ \int_{\mathbb{R}} E\left[(r_{T'}(\xi_T) - r(\xi_T))^2 \mathbf{I}_{\xi_T \in \Delta} \mid \xi_T = x\right] dP_{\xi_T}(x)
\end{aligned}$$

thus

$$D_T = \int_\Delta E\left[(r_{T'}(x) - r(x))^2 \mid \xi_T = x\right] f(x)dx \ .$$

Now Lemma 3.1 entails

$$\begin{aligned}
(5.40) \qquad D_T(x) &= \ : E\left[(r_{T'}(x) - r(x))^2 \mid \xi_T = x\right] \\
&\le \ E\left[r_{T'}(x) - r(x)\right]^2 + 8\sup_{x \in \Delta} |m(x)|\varphi_{rev}(T - T') \ ,
\end{aligned}$$

$x \in \Delta$, except on a P_{ξ_0}-null set.

Consequently

$$(5.41) \qquad D_T(x) \le \sup_{x \in \Delta} E[r_{T'}(x) - r(x)]^2 + 8\sup_{x \in \Delta} |m(x)|a\rho^{(T-T')} \ .$$

Using the bound (5.41) in (5.40) we get

$$D_T \le \sup_{x \in \Delta} E[r_{T'}(x) - r(x)]^2 + 8\sup_{x \in \Delta} |m(x)|a\rho^{(T-T')} \ .$$

Now from I_2 we may obtain uniform majorizations as in the proof of Theorem 5.2, hence

$$
\begin{aligned}
D_T &= O\left(\frac{1}{T'}\right) + O(\rho^{(T-T')}) \\
&= O\left(\frac{1}{T}\right)
\end{aligned}
$$

which proves (5.39). ■

Note that the **nonparametric predictor** ζ^*_{T+H} reaches a **parametric rate**. Once more this fact proves the value of nonparametric methods.

Notes

All the results in this chapter are new or very recent.

N. CHEZE-PAYAUD (1994) has proved Corollaries 5.1 and 5.2, and Theorems 5.3 and 5.5. Results about quadratic error associated with admissible sampling may be found in [CP].

The other results have been obtained by the author of the present work.

APPENDIX

NUMERICAL RESULTS

The following tables and figures compare numerical performances of non-parametric predictors (N.P.) and parametric predictors (B.J.).

k is chosen by using (3, 60), $K(x) = (2\pi)^{-\frac{k}{2}} e^{-\frac{\| x \|^2}{2}}$, $x \in \mathbb{R}^k$ and $h_n = \widehat{\sigma}_n \ n^{-1/(4+k)}$ where $\widehat{\sigma}_n = \left(\dfrac{1}{n} \displaystyle\sum_{t=1}^{n} (\xi_t - \overline{\xi}_n)^2 \right)$; $\xi_1, \dots, \xi_n$ are the data and $\overline{\xi}_n = \dfrac{1}{n} \displaystyle\sum_{t=1}^{n} \xi_t$. The horizon is 1.

In order to measure the prediction error two criteria are used : the EMO and the EMP defined as

$$EMO = \frac{1}{p} \sum_{j=n-p+1}^{n} \left| \frac{\xi_j - \widehat{\xi}_j}{\xi_j} \right|$$

and

$$EMP = \frac{1}{p} \sum_{j=n-p+1}^{n} \left| \frac{\widehat{q}_j}{\widehat{\xi}_j} \right|$$

where $\widehat{\xi}_j$ is the prediction of ξ_j constructed from the data $\xi_1, \dots, \xi_{j-1}$ and $\widehat{q}_j$ the empirical quantile associated with the theoretical quantile q_j defined by

$$P(|\widehat{\xi}_j - \xi_j| < q_j) = 0.95 .$$

Series 1 to 17 are taken from CARBON-DELECROIX [C-D] paper.

Series 1 to 9 are obtained by simulation, series 10 to 17 are extracted from [B-J], [K-S] and [P]. In each case the best predictor is marked with a star. For the EMO the nonparametric predictor is better than B.J. 12 times out of 17 and for the EMP 14 times out of 17.

Series 18 shows the good behaviour of N.P. when the horizon increases (cf. table 18, figures 2 and 3). It deals with french cars registrations from april 1987 to september 1988.

Series 19 and 20 are extracted from the paper of POGGI ([PO]) devoted to nonparametric prediction of global french electricity consumption. The results are very good!

AR1 $x_t = 0.9x_{t-1} + 1000 + \varepsilon_t$ $\varepsilon_t \sim> N(0,5)$ $n = 100, \ H = 5$		
B.J. (p,d,q)	*EMO*	*EMP*
$(1,0,0)$	0.089*	0.136*
$(0,0,2)$	0.135	0.141
$(1,0,1)$	0.089*	0.140*
N.P. p		
1	0.098	0.128
5	0.077	0.120
10	0.085	**0.116***
$\hat{p} = 19$	**0.062***	0.126

Table 1

AR1 (limit) $x_t = 0.99x_{t-1} + 1000 + \varepsilon_t$ $\varepsilon_t \sim> N(0,5)$ $n = 100, \ H = 5$		
B.J. (p,d,q)	*EMO*	*EMP*
$(0,1,1)$	0.014	0.022*
$(0,1,2)$	0.013	0.022*
$(0,1,3)$	0.012*	0.026
N.P. $(d=1)p$		
1	0.011	0.023
2	0.007	0.023
$\hat{p} = 6$	**0.003***	0.023
10	0.007	**0.019***

Table 2

MA6 $x_t = \varepsilon_t - 2.848\varepsilon_{t-1} + 2.6885\varepsilon_{t-2} - 1.64645\varepsilon_{t-3}$
$+2.972\varepsilon_{t-4} - 2.1492\varepsilon_{t-5} + 0.67716\varepsilon_{t-6}\varepsilon_t \sim > N(0,5)$
$n = 100, \ H = 5$

B.J. (p,d,q)	EMO	EMP
$(0,0,6)$	3.03	5.33
$(0,0,7)$	2.78*	5.31
$(1,0,2)$	3.01	5.30*
N.P. p		
1	3.03	5.28
$\hat{p} = 2$	**2.77***	**5.16***
5	3.14	5.62
10	4.44	6.18

Table 3

AR2 $x_t = 0.7x_{t-1} + 0.2x_{t-2} + 1000 + \varepsilon_t \ \ \varepsilon_t \sim > N(0,5)$
$n = 100, \ H = 5$

B.J. (p,d,q)	EMO	EMP
$(2,0,0)$	**0.012***	0.138*
$(1,0,0)$	0.026	0.154
$(3,0,0)$	0.013	0.144
$(0,1,3)$	0.026	0.148
N.P. p		
1	0.019	0.136
2	0.014*	0.143
10	0.024	0.146
$\hat{p} = 30$	**0.015**	**0.074***

Table 4

ARMA(1,1) $x_t = 0.8x_{t-1} + \varepsilon_t + 0.2\varepsilon_{t-1} + 1000$ $\varepsilon_t \sim> N(0,5)\ \ n = 100,\ \ H = 5$		
B.J. (p, d, q)	*EMO*	*EMP*
$(2, 0, 0)$	0.177	0.294
$(1, 0, 0)$	0.149	0.282*
$(1, 0, 1)$	0.123*	0.290
$(1, 0, 2)$	0.170	0.296
N.P. p		
5	0.149	0.326
10	0.098	0.313
20	0.099	0.316
$\hat{p} = 30$	**0.074***	**0.186***

Table 5

AR1 $x_t = 0.8x_{t-1} + 1000 + \varepsilon_t$ $\varepsilon_t \sim> \exp(1/300)\ n = 100,\ H = 5$		
B.J. (p, d, q)	*EMO*	*EMP*
$(1, 0, 0)$	**1.60***	11.0*
$(0, 0, 2)$	3.75	11.7
$(0, 0, 1)$	4.23	11.5
$(2, 0, 0)$	**1.60***	11.3
N.P. p		
1	2.45	9.65
5	1.77*	12.52
$\hat{p} = 7$	4.92	12.42
30	2.55	**6.55***

Table 6

AR1 (contaminated) $y_t = 0.5(1 - \delta_t)y_{t-1} + (1 - 3\delta_t'/4)\varepsilon_t$ $P(\delta_t = 1) = P(\delta_t = 0) = 1/2,\ P(\delta_t' = 0) = 2/3\ P(\delta_t' = 1) = 1/3$ $\varepsilon_t \sim> N(0,1)\ \ n = 100,\ H = 5$		
B.J. (p, d, q)	EMO	EMP
$(1, 0, 0)$	153.8	218.0
$(3, 0, 0)$	152.5	214.0
$(7, 0, 0)$	146.5	213.5
$(10, 0, 0)$	137.8*	**198.8***
N.P. p		
5	80.1	272.2
$\hat{p} = 10$	**51.9***	219.4*
20	86.6	288.5
30	64.5	320.6

Table 7

AR1 (contaminated) $z_t = y_t + 100$		
B.J. (p, d, q)	EMO	EMP
$(1, 0, 0)$	1.63	3.57*
$(0, 0, 3)$	**1.49***	3.58
$(7, 0, 0)$	1.82	3.61
$(0, 0, 1)$	1.62	3.57*
N.P. p		
5	1.99	**3.54***
$\hat{p} = 10$	1.71	4.12
20	1.69*	6.41
30	1.85	4.51

Table 8

Perturbated sinusoïd $x_t = 3000\sin(\pi t/15) + \varepsilon_t$ $\varepsilon_t \sim \exp(1/300)$ $n = 200,\ H = 5$		
B.J. $(p,d,q)(P,D,Q)^T$	*EMO*	*EMP*
$(2,1,0)(2,1,0)^{30}$	21.88*	47.10
$(2,1,1)(2,1,1)^{30}$	25.08	40.70*
N.P. p		
$\hat{p} = 15$	**7.81***	32.76
30	9.82	**29.50***
60	13.33	34.35

Table 9

Profit margin (A. Pankratzz)$SARIMA$ $n = 80,\ H = 5$		
B.J. $(p,d,q)(P,D,Q)^T$	*EMO*	*EMP*
$(1,0,0)(2,1,4)^4$	4.85*	24.10*
N.P. p		
4	8.40	17.41
8	6.92	16.89
12	4.98	17.23
$\hat{p} = 24$	**1.17***	**9.19***

Table 10

Cigar consumption (A. Pankratzz) $SARIMA$ $n = 96, \; H = 6$		
B.J. $(p,d,q)(P,D,Q)^T$	EMO	EMP
$(1,1,0)(1,2,0)^{12}$	13.07	42.70
$(2,1,0)(1,1,0)^{12}$	8.76*	**23.1***
N.P. p		
4	12.26	32.73
$\hat{p} = 12$	**5.70***	24.95
24	7.83	24.63*

Table 11

Change in business inventories (A. Pankratzz) $n = 60, \; H = 10$		
B.J. (p,d,q)	EMO	EMP
$(1,0,0)$	37.0	156.9
$(2,0,0)$	36.6*	156.5*
$(3,0,0)$	39.1	172.3
N.P. p		
$\hat{p} = 1$	65.7	165.0
10	**28.8***	81.5
20	32.8	**59.4***

Table 12

Coal (A. Pankratzz) $n = 90$, $H = 10$		
B.J. (p, d, q)	EMO	EMP
$(1, 0, 0)$	3.83	23.60*
$(2, 0, 0)$	3.42	24.20
$(1, 0, 1)$	3.52	23.90
$(1, 0, 2)$	3.47	24.32
$(1, 0, 3)$	3.11*	24.06
N.P. p		
1	**2.94***	19.53
2	3.14	19.84
$\hat{p} = 3$	3.22	**19.39***
5	4.04	19.63
10	3.50	22.51

Table 13

Yields from a batch chemical process (G. Box, G. Jenkins) $n = 70$, $H = 5$		
B.J. (p, d, q)	EMO	EMP
$(1, 0, 1)$	26.75	**42.90***
$(2, 0, 0)$	26.31	43.14
$(0, 0, 1)$	26.26	43.19
$(0, 0, 2)$	25.70*	43.02
N.P. p		
$\hat{p} = 2$	**17, 88***	44.38*
5	23.10	47.12
10	29.88	55.13
20	35.01	50.18

Table 14

Chemical process concentration readings (G. Box, G. Jenkins) $n = 197$, $H = 10$		
B.J. (p, d, q)	*EMO*	*EMP*
$(1, 0, 1)$	2.48	4.01
$(1, 0, 2)$	2.38	3.96*
$(0, 1, 1)$	**1.85***	4.17
N.P. (p, d)		
$(\hat{p} = 2, 0)$	2.72	**3.89***
$(5, 0)$	2.71	4.06
$(\hat{p} = 1, 1)$	2.11	4.33
$(5, 1)$	3.07	4.55
$(10, 1)$	1.91*	4.63

Table 15

AR1 $x_t = 0.9 x_{t-1} + \varepsilon_t$ $\varepsilon_t \sim>$ uniform on $[-49, 49]$ (M. Kendall, A. Stuart) $n = 100$, $H = 5$		
B.J. (p, d, q)	*EMO*	*EMP*
$(1, 0, 1)$	41.8	286*
$(0, 0, 2)$	**30.6***	365
$(0, 0, 3)$	43.8	333
$(0, 0, 4)$	36.6	343
N.P. p		
1	137.2	445
5	456.0	1274
10	63.5*	**47***
$\hat{p} = 18$	312	869

Table 16

AR3 $x_t = 0.2x_{t-1} + x_{t-2} - 0.3x_{t-3} + \varepsilon_t$ $\varepsilon_t \sim>$ uniform on $[-49, 49]$ (M. Kendall, A. Stuart) $n = 100,\ H = 5$		
B.J. (p, d, q)	EMO	EMP
$(1, 0, 0)$	52.9	197*
$(0, 0, 3)$	44.6*	462
$(1, 0, 2)$	95.9	207
$(2, 0, 0)$	53.2	222
$(3, 0, 0)$	70.1	**204**
N.P. p		
1	115.7	316
5	137.3	362
10	529.2	630
20	**44.5***	**66***
$\hat{p} = 25$	49.0	**66***

Table 17

French car registrations (april 1987 - september 1988)

t	x_t	$\hat{x}_t BJ$	$\hat{x}_t NP\ \hat{k} = 36$
1	192.1	183.1	**197.1**
2	156.7	**173.8**	179.1
3	151.2	**170.5**	180.7
4	195.9	161.9	**167.9**
5	146.1	136.3	**138.7**
6	129.6	**134.4**	144.1
7	232.3	189.1	**195.9**
8	197.8	190.0	**192.5**
9	208.9	193.9	**204.6**
10	160.6	148.6	**156.3**
11	160.0	153.9	**164.5**
12	218.0	206.2	**221.3**
13	189.0	**188.6**	197.1
14	184.0	181.3	**185.6**
15	141.6	**179.8**	195.5
16	210.0	**163.6**	160.3
17	157.1	133.9	**135.4**
18	146.7	134.4	**141.1**

Table 18

−− data
— predictions BJ
... predictions NP

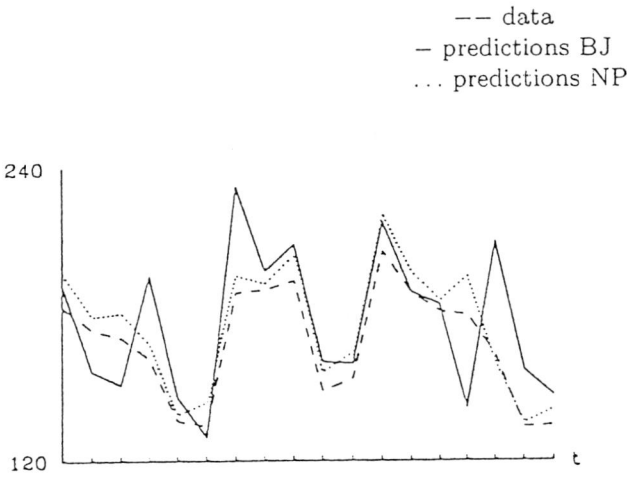

French car registrations
Figure 2

— errors BJ
− − − errors NP

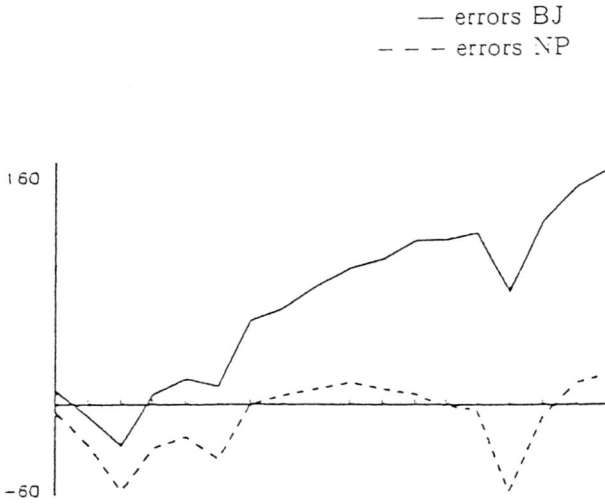

French car registrations (cumulated prediction errors)
Figure 3

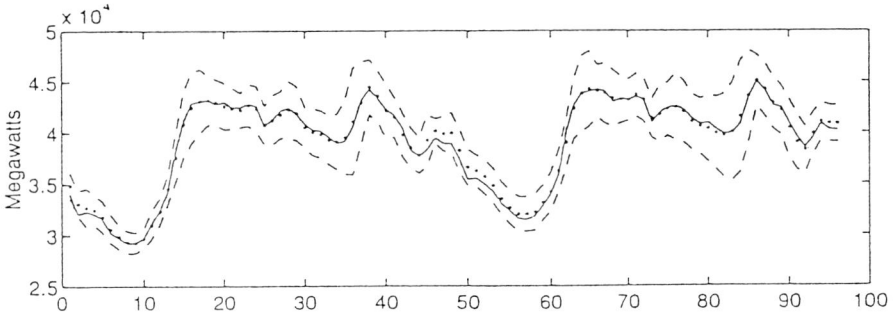

— data
... predictions
— — — ± 3 standard deviation

Prediction : 4th and 5th february 1991
Relative error : 0.8986 % French electricity consumption
Figure 4

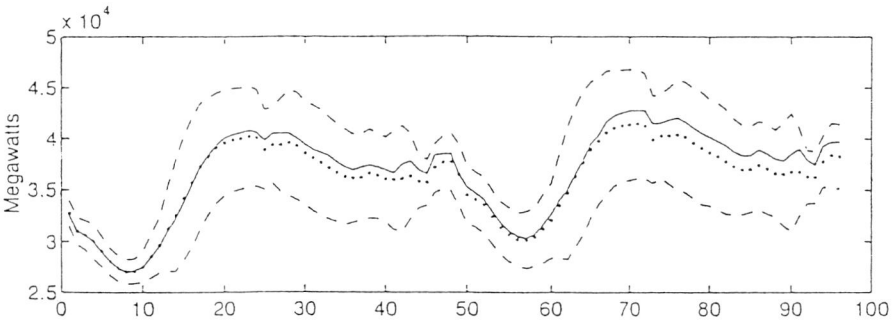

— data
... predictions
— — — ± 3 standard deviation

Prediction : 26th and 27th august 1991
Relative error : 2.159 % French electricity consumption
Figure 5

BIBLIOGRAPHY

References of Synopsis

[A-O] ANTONIADIS A. - OPPENHEIM G. (1995) editors Wavelets and statistics. Springer Verlag.

[A-G] ASH R.B. - GARDNER M.F. (1975). Topics in stochastic processes. Academic Press.

[B-PR] BASAWA I.V. - PRAKASA RAO B.L.S. (1980). Statistical inference for stochastic processes. Academic Press.

[BI-MA] BIRGE L. - MASSART P. (1995). From model selection to adaptive estimation. Preprint 95.41 Univ. Paris Sud.

[BO] BOSQ D. (1991). Modelization, Nonparametric estimation and prediction for continuous time processes in "Nonparametric functional estimation and related topics". Ed. G. ROUSSAS - NATO ASI Series, 509-529.

[BO-LE] BOSQ D. - LECOUTRE J.P. (1987). Théorie de l'estimation fonctionnelle. Economica.

[BX-JE] BOX G. - JENKINS G. (1970). Time series analysis : forecasting and control. Holden Day.

[BR-DA] BROCKWELL P.J. - DAVIS R.A. (1991). Time series : theory and methods. Springer Verlag.

[CA-DE] CARBON M. - DELECROIX M. (1993). Nonparametric forecasting in time series : a computational point of view. Applied Stoch. Models and data analysis 9, 3, 215-229.

[DV] DEVROYE, L. (1987). A course in density estimation. Birkhäuser.

[DV-GY] DEVROYE L. - GYORFI L. (1985). Nonparametric density estimation. The $L1$ view. Wiley.

[GO-MO] GOURIEROUX C. - MONFORT A. (1983). Cours de série temporelle. Economica.

[GR] GRENANDER U. (1981). Abstract inference. Wiley.

[KO-TS] KOROSTELEV A.P. - TSYBAKOV A.B. (1993). Minimax theory of image reconstruction. Springer Verlag.

[MI] MILCAMPS B. (1995). Analyse des corrélations inter-marché. Mémoire ISUP (Paris).

[NA] NADARAJA E.A. (1964). On estimating regression. Theory Probab. An., 141-142.

[PA] PARZEN E. (1962). On the estimation of probability density function and mode. Ann. Math. Statist. 33, 1065-1076.

[PO] POGGI J.M. (1994). Prévision non paramétrique de la consommation électrique. Rev. Statist. Appliq. XLII (4), 83-98.

[PR] PRAKASA RAO B.L.S. (1983). Nonparametric functional estimation. Academic Press.

[RB-ST] ROBINSON P.M. - STOYANOV J.M. (1991). Semiparametric and nonparametric inference from irregular observations on continuous time stochastic processes in "Nonparametric functional estimation and related topics". Ed. G. ROUSSAS NATO ASI Series, 553-558.

[RO1] ROSENBLATT M. (1956). Remarks on some non parametric estimates of a density function. Ann. Math. Statist. 27, 832-837.

[RO2] ROSENBLATT M. (1956). A central limit theorem and a strong mixing condition. Proc. Nat. Ac. Sc. USA 42, 43-47.

[WA] WATSON G.S. (1964). Smooth regression analysis. Sankhya, A, 26, 359-372.

[Y] YAKOWITZ S. (1985). Markov flow models and the flood warning problem. Water resources research, 21, 81-88.

References of Chapter 1

[BE] BENNETT G. (1962). Probability inequalities for sum of independent random variables. J. Amer. Statis. Assoc. 57, 33-45

[BER] BERBEE H.C.P.(1979) Random walks with stationary increments and renewal theory. Math. Centr. Tract. Amsterdam.

[BI1] BILLINGSLEY P. (1968). Convergence of Probability Measures. Wiley

[BI2] BILLINGSLEY P. (1986). Probability and Measure. (2nd edition) Wiley

[BO1] BOSQ D. (1993). Bernstein-type large deviation inequalities for partial sums of strong mixing processs. Statistics,24, 59-70.

[BO2] BOSQ D. (1995). Optimal asymptotic quadratic error of density estimators for strong mixing or chaotic data. Statistics and Probability Letters. 22, 339-347.

[BR1] BRADLEY R. (1983). Approximation theorems for strongly mixing random variables. Michigan Maths. J. 30, 69-81.

[BR2] BRADLEY R. (1986). Basic Properties of strong mixing conditions. in E.Eberlein and M.S. Taqqu editors, Dependence in Probability and Statistics, p.165-192. Birkhäuser.

[CA] CARBON M. (1993). Une nouvelle inégalité de grandes déviations. Applications. Publ. IRMA, Vol. 32 n° 11.

[DA] DAVYDOV Y.A. (1968). Convergence of distributions generated by stationary stochastic Processes. Theor. Probab. Appl. 13, 691-696.

[DO] DOOB J.L. (1967). Stochastic Processes. (7th printing). Wiley

[DK] DOUKHAN P. (1994). Mixing -Properties and Examples . Lecture Notes in Statistics. Springer Verlag.

[HO] HOEFFDING (1963). Probability inequalities for sums of bounded random variables. J. Amer. Statist. Assoc. 58, 13-30.

[IB] IBRAGIMOV I.A. (1962). Some limits theorems for stationary Processes. Theor. Prob. Appl. 7, 349-382.

[P-T] PHAM D.T.-TRAN L.T. (1985). Some strong mixing properties of time series models. Stoch. Proc. Appl. 19, 207-303.

[RH] RHOMARI N. (1994). Filtrage non paramétrique pour les processus non Markoviens. Applications. Ph D. Thesis University Pierre et Marie Curie (Paris).

[RI] RIO E. (1993). Covariance inequalities for strongly mixing processes. Ann. Inst. Henri Poincaré, 29, 4, 587-597.

[RO] ROSENBLATT M. (1956). A central limit theorem and a strong mixing condition. Proc. Nat. Ac. Sc. USA, 42, 43-47.

[ST] STOUT W.F. (1974). Almost sure convergence. Academic Press.

[TR] TRAN L.T. (1989). The L^1 convergence of kernel density estimates under dependence. The Canad. J. of Statist. 17, 2, 197-208.

[YO] YOKOYAMA R. (1980). Moments bound for stationary mixing sequences. Z. Wahrsch. Gebiete, 52, 45-57.

References of Chapter 2

[BI1] BILLINGSLEY P. (1968) Convergence of Probability Measures. Wiley.

[BO1] BOSQ D. (1989) Non parametric estimation of a Non Linear Filter using a Density Estimator with zero-one "explosive" behaviour in $\mathbb{R}^d$. Statistics and Decisions 7, 229 − 241.

[BO2] BOSQ D. (1995) Optimal Asymptotic quadratic Error of Density Estimators for strong mixing or chaotic data. Statistics and Probab. Letters 22, 339-347.

[EP] EPANECHNIKOV V.A. (1969) Nonparametric estimation of a multidimensional probability density. Theory Probab. Appl. 14, 153-158.

[FA1] FAN (1991) On the optimal rates of convergence for nonparametric deconvolution problems. Annals of Statist., 19, 1257-1272.

[FA2] FAN (1991) Asymptotic normality for deconvolving kernel density estimators, Sankhya, Ser. A 53, 97-110.

[FA3] FAN (1992) Deconvolution with supersmooth distributions. Canad J. Statist. 20, 155-169.

[LM] LASOTA A. and MACKEY M.C. (1985) Probabilistic Properties of deterministic systems. Cambridge Univ. Press.

[MA1] MASRY E. (1986) Recursive Probability Density Estimation for weakly Dependent Stationary processes. IEEE Trans. inform. Theory. II, 32, 2, 254-267.

[MA2] MASRY E. (1991) Multivariate Probability Density deconvolution for stationary random processes. IEEE Trans. inform. Theory II, 37, 1105-1115.

[MA3] MASRY E. (1993) Strong consistency and rates for deconvolution of multivariate densities of stationary processes. Stoch. Proc. and Applic. 47, 53- 74.

[MA4] MASRY E. (1993) Asymptotic Normality for deconvolution estimators of multivariate densities of stationary processes. J. Multivariate Analysis, 44, 47- 68.

[PA] PARZEN E. (1962) On the estimation of a Probability density function and mode. Ann. Math. Statist. 33, 1065-1076.

[PT] PHAM T.D. and TRAN L.T. (1991) Kernel Density Estimation under a locally Mixing condition - in Nonparametric Functional Estimation and Related Topics. Ed. G. ROUSSAS NATO ASI SERIES V. 335, 419-430.

[RB] ROBINSON P.M. (1983) Nonparametric Estimators for time series. J. Time Series Anal. 4, 185-297.

[RO2] ROSENBLATT M. (1956) Remarks on some nonparametric estimates of a density function. Ann. Math. Statist. 27, 832-837.

[RO3] ROSENBLATT M. (1985) Stationary Sequences and Random fields. Birkhäuser.

[RS1] ROUSSAS G. (1967) Nonparametric Estimation in Markov Processes. Ann. Inst. Stat. Math. 21, 73-87.

[RS2] ROUSSAS G. (1988) Nonparametric Estimation in mixing sequences of random variables. J. Stat. Plan. and Inf. 18, 135-149.

[RU] RUELLE D. (1989) Chaotic Evolution and Strange Attractors. Cambridge Univ. Press.

[ST] STUTE W. (1982) A law of the logarithm for kernel density estimators. Ann. Probab. 10, 414-422.

[TR] TRAN L.T. (1990) Kernel density and regression estimation for dependent random variables and time series. Techn. report. Univ. Indiana.

[TGS] TRUONG Y.K. and STONE C.J. (1992) Nonparametric function estimation involving time series. Ann. Statist. 20, 77-98.

References of Chapter 3

[AN-PO] ANGO NZE P. and PORTIER B. (1994) Estimation of the density and of the regression functions of an absolutely regular stationary process. Publ. ISUP 38, 59-87.

[BE-DE] BERLINET A. and DEVROYE L. (1994) A comparison of kernel density estimates. Publ. ISUP Vol. 38, 3, 3-59.

[BE-FR] BOENTE G. and FRAIMAN R. (1995) Asymptotic distribution of data driven smoothers in density and regression under dependence. The Canadian Journ. of Statist. 23,4, 383-397.

[BO] BOSQ D. (1991) Nonparametric prediction for unbounded almost stationary processes. Nonparametric functional estimation and related topics. NATO ASI series Vol. 335 - ed. G. ROUSSAS. 389-404.

[BO-CH] BOSQ D. and CHEZE-PAYAUD N. (1995) Optimal asymptotic quadratic error of nonparametric regression function estimates for a continuous-time process from sampled data. To appear.

[BO-JE] BOX G. and JENKINS G. (1970) Time series analysis : forecasting and control. Holden-Day.

[BR-DA], BROCKWELL P.J. and DAVIS R.A. (1991) Time series : theory and methods. Springer-Verlag.

[BRO] BRONIATOWSKI M. (1993) Cross validation methods in kernel nonparametric density estimation : a survey. Publ. ISUP, 38, 3-4, 3-28.

[CA-DE] CARBON M. and DELECROIX M. (1993) Nonparametric forecasting in time series, a computational point of view. Applied Stoch. Models and data Analysis, 9, 3, 215-229.

[CO] COLLOMB G. (1981) Estimation non paramétrique de la régression. Revue bibliographique. Ins. Statist. Rev. 49, 75-93.

[DE-HO] DEHEUVELS P. and HOMINAL P. (1980) Estimation automatique de la densité. Rev. Statist. Appl. 28, 25-55.

[DC-OP-TH] DACUNHA-CASTELLE D., THOMASSONE R. and OPPENHEIM G.(1995) Prévision des pointes d'ozone à Paris (preprint).

[DE] DELECROIX M. (1987) Sur l'estimation et la prévision non paramétrique des processus ergodiques. Thèse de Doctorat d'Etat, Université de Lille.

[DE-RO] DELECROIX M. and ROSA A.C. (1995) Ergodic processes prediction via estimation of the conditional distribution function. Publ. ISUP, XXXIX, 2, 35-56.

[DO] DOUKHAN P. (1991) Consistency of delta-sequence estimates of a density or a regression function for a weakly stationary sequence. Séminaire Univ. Orsay 1989-90, 121-141.

[EP] EPANECHNIKOV V.A. (1969) Nonparametric estimation of a multidimensional probability density. Theory Probab. Anl. 14, 153-158.

[GO-MO] GOURIEROUX C. and MONFORT A. (1983) Cours de séries temporelles. Economica. Paris.

[GE] GUERRE E. (1995) The General behavior of Estimators of the Box-Cox model for integrated Time Series. Preprint INSEE.

[HA-1] HÄRDLE W. (1990) Applied nonparametric regression. Cambridge University Press.

[HA-2] HÄRDLE W. (1991) Smoothing Techniques with implementation in S. Springer Verlag.

[MAE] MAES J. (1994) Estimation non paramétrique de la fonction chaotique et des exposants de Lyapunov d'un système dynamique. C.R. Acad., Sci. Paris t. 319 ser. 1 p. 1005-1008.

[MR] MARRON J.J. (1991) Root n bandwith selection. Nonparametric functional estimation and related topics. NATO ASI series, Vol 335, ed. G. ROUSSAS, 251-260.

[MS] MASRY E. (1994) Multivariate regression estimation with errors in variables for stationary processes. Nonparametric Statist. 3 in press.

[NA] NADARAJA E.A. (1964) On estimating regression. Theory Probab. An. 9, 141- 142.

[PO] POGGI J.M. (1994) Prévision non paramétrique de la consommation électrique. Rev. Stat. Appl. XLII, 4, 83-98.

[RH] RHOMARI N. (1994) Filtrage non paramétrique pour les processus non Markoviens. Application. Thèse Doctorat Univ. Paris 6.

[RO] ROBINSON P.M. (1983) Nonparametric estimators for time series. J. Time Ser. Anal. 4(3), 185-207.

[RO] ROSENBLATT M. (1985) Stationary Sequences and Random Fields. Birkhäuser.

[RS] ROUSSAS G.G. (1990) Nonparametric regression estimation under mixing conditions. Stoch. Processes Appl. 36, 107-116.

[RS-IO] ROUSSAS G.G. - IOANNIDES D. (1987) Moment inequalities for mixing sequences of random variables. Stochastic Anal. Appl. 5(1), 61-120.

[SA] SARDA P. (1993) Smoothing parameter selection for smooth distribution function. J. Statist. Plan. Infer 35, 65-75.

[SI] SILVERMAN B.W. (1986) Density estimation for Statistics and Data Analysis. Chapman and Hall.

[ST-TR] STONE C.J. - TRUONG Y.K. (1992) Nonparametric function estimation involving time series, Annals of Statist., 20, 1, 77-97.

[TR] TRAN L.T. (1993) Nonparametric function for time series by local average estimators. Ann. Statist. 21(2), 1040-1057.

[VI] VIEU P. (1994) Quelques résultats en estimation fonctionnelle. Mémoire d'Habilitation, Univ. P. Sabatier, Toulouse (France).

[WA] WATSON G.S. (1964) Smooth regression analysis. Sankhya Ser. A, 26, 359-372.

References of Chapter 4

[AD] ADLER R.J. (1990) An introduction to continuity, extrema, and Related topics for general Gaussian processes. Inst. of Math. Statist., Hayward, California.

[BA] BANON G. (1978) Nonparametric identification for diffusion processes. Siam J. Control and Optimisation V16, 380-395.

[BA-NG1] BANON G. and NGUYEN H.T. (1978) Sur l'estimation récurrente de la densité et de sa dérivée pour un processus de Markov, C.R. Acad. Sci. Paris t. 286, sér. A, 691-694.

[BA-NG2] BANON G. and NGUYEN H.T. (1981). Recursive estimation in diffusion model. Siam J. Control and optimisation V10, 676-685.

[BI] BILLINGSLEY P. (1986). Probability and measure. Wiley.

[BO1] BOSQ D. (1993) Optimal and superoptimal quadratic error of functional estimators for continuous time processes. Preprint, Univ. Paris VI.

[BO2] BOSQ D. (1995) Sur le comportement exotique de l'estimateur à noyau de la densité marginale d'un processus à temps continu. C.R. Acad. Sci. Paris t. 320, sér.I, 369-372.

[CA-LE] CASTELLANA J.V. and LEADBETTER M.R. (1986) On smoothed Probability density estimation for stationary processes. Stoch. Proc. Appl., 21, 179-193.

[DE] DELECROIX M. (1980) Sur l'estimation des densités d'un processus stationnaire à temps continu. Publ. ISUP, XXV, 1-2, 17-39.

[FA] FARREL R. (1972) On the best obtainable asymptotic rates of convergence in estimation of a density function at a point. Ann. of Math. Stat. 43, 1, 170-180.

[IB-HA] IBRAGIMOV I.A. and HASMINSKII R.Z. (1981) Statistical estimation - Asymptotic theory. Springer-Verlag, New-York.

[IB-RZ] IBRAGIMOV I.A. and ROZANOV Y.A. (1978) Gaussian random processes. Springer Verlag, New-York.

[KU] KUTOYANTS Y.A. (1995) On density estimation by the observations of ergodic diffusion process. Preprint Un. du Maine.

[LE] LEBLANC F. (1995) PhD thesis, Univ. Paris VI.

[MA1] MASRY E. (1983) Probability density estimation from sampled data. IEEE transf. inf. th. 29, 696-709.

[MA2] MASRY E. (1988) Continuous-parameter stationary process : Statistical properties of joint density estimators, J. of mult. Anal., 26, 133-165.

[NG] NGUYEN H.T. (1979) Density estimation in a continuous-time stationary Markov process, Annals of Statist., 7, 2, 341-348.

[NG-PH1] NGUYEN H.T. and PHAM D.T. (1980) Sur l'utilisation du temps local en Statistique des processus, C.R. Acad. Sci. Paris, 290, A, 165-170.

[NG-PH2] NGUYEN H.T. and PHAM D.T. (1981) Nonparametric estimation in diffusion model by discrete sampling.

[PR1] PRASAKA RAO B.L.S. (1979) Nonparametric estimation for continuous time Markov processes via delta-families. Publ. ISUP, XXIV, 81-97.

[PR2] PRASAKA RAO B.L.S. (1990) Nonparametric density estimation for stochastic process from sampled data. Publ. ISUP, XXXV, 51-84.

[RA] RAO (1992) Probability Theory. Acad. Press.

References of Chapter 5

[DB] BOSQ D. (1993) - Vitesses optimales et superoptimales des estimateurs fonctionnels pour un processus à temps continu. C.R. Acad. Sci. Paris 317, ser. I, 1075-1078.

[CP] CHEZE-PAYAUD N. (1994) - Régression, prédiction et discrétisation des processus à temps continu. Thesis Univ. Paris 6.

References of appendix

[B-J] BOX G. - JENKINS G. (1970). Times series analysis forecasting and control. Holden-Day.

[C-D] CARBON M. - DELECROIX M. (1993). Nonparametric forecasting in time series : a computational point of view. Applied stoch. Models and data Analysis 9, 3, 225 − 229.

[K-S] KENDALL M. - STUART A. (1976). The advanced theory of Statistics. Vol. 3 - C. Griffin and Co., third edition.

[P] PANKRATZ A. (1983). Forecasting with univariate Box-Jenkins models. Concepts and cases. Wiley.

[PO] POGGI J.M. (1994). Prévision non paramétrique de la consommation électrique. Rev. Statist. Appl. XLII, 4, 83 − 98.

Index

A

Absolute regularity 16
Adaptive method 14
Admissible sampling 13, 122, 139
α-mixing 7, 16
ARMA process 1, 92
Asymptotic normality 9, 11, 34, 52, 73, 78, 95, 137
Autoregressive processes (infinite dimensional) 14

B

β-mixing 16
Berbee's lemma 17
Bernstein's inequality 22
Billingley's inequality 20
Bochner's lemma 42, 43
Borel-Cantelli lemma in continuous time 111
Box-Cox transformation 85
Box-Jenkins (method) 1, 92
Bradley's lemma 18

C

Cadlag 96, 121
Cars registrations 144, 153, 155
Central limit theorem 34
Chaos 84
Chaotic data 55
Coupling 17
Covariance inequalities 18, 19
Cramer's conditions 22
Cross validation 90

D

Davydov's inequality 20
Density kernel estimator 3, 40, 95
Deseasonalization 14, 86-87
Dichotomy 119, 139
Differencing 12, 87
Differentiable sample paths 108
Diffusion process 105
Double kernel method 91
Dynamical system 56

E

F

G

H

I

K

L

Lecture Notes in Statistics

For information about Volumes 1 to 23
please contact Springer-Verlag

Vol. 24: T.S. Rao, M.M. Gabr, An Introduction to Bispectral Analysis and Bilinear Time Series Models. viii, 280 pages, 1984.

Vol. 25: E. Parzen (Editor), Time Series Analysis of Irregularly Observed Data. Proceedings, 1983. vii, 363 pages, 1984.

Vol. 26: J. Franke, W. Härdle and D. Martin (Editors), Robust and Nonlinear Time Series Analysis. Proceedings, 1983. ix, 286 pages, 1984.

Vol. 27: A. Janssen, H. Milbrodt, H. Strasser, Infinitely Divisible Statistical Experiments. vi, 163 pages, 1985.

Vol. 28: S. Amari, Differential-Geometrical Methods in Statistics. v, 290 pages, 1985.

Vol. 29: B.J.T. Morgan and P.M. North (Editors), Statistics in Ornithology. xxv, 418 pages, 1985.

Vol 30: J. Grandell, Stochastic Models of Air Pollutant Concentration. v, 110 pages, 1985.

Vol. 31: J. Pfanzagl, Asymptotic Expansions for General Statistical Models. vii, 505 pages, 1985.

Vol. 32: R. Gilchrist, B. Francis and J. Whittaker (Editors), Generalized Linear Models. Proceedings, 1985. vi, 178 pages, 1985.

Vol. 33: M. Csörgo, S. Csörgo, L. Horváth, An Asymptotic Theory for Empirical Reliability and Concentration Processes. v, 171 pages, 1986.

Vol. 34: D.E. Critchlow, Metric Methods for Analyzing Partially Ranked Data. x, 216 pages, 1985.

Vol. 35: T. Calinski and W. Klonecki (Editors), Linear Statistical Inference. Proceedings, 1984. vi, 318 pages, 1985.

Vol. 36: B. Matérn, Spatial Variation. Second Edition. 151 pages, 1986.

Vol. 37: R. Dykstra, T. Robertson and F.T. Wright (Editors), Advances in Order Restricted Statistical Inference. Proceedings, 1985. viii, 295 pages, 1986.

Vol. 38: R.W. Pearson and R.F. Boruch (Editors), Survey Research Designs: Towards a Better Understanding of Their Costs and Benefits. v, 129 pages, 1986.

Vol. 39: J.D. Malley, Optimal Unbiased Estimation of Variance Components. ix, 146 pages, 1986.

Vol. 40: H.R. Lerche, Boundary Crossing of Brownian Motion. v, 142 pages, 1986.

Vol. 41: F. Baccelli, P. Brémaud, Palm Probabilities and Stationary Queues. vii, 106 pages, 1987.

Vol. 42: S. Kullback, J.C. Keegel, J.H. Kullback, Topics in Statistical Information Theory. ix, 158 pages, 1987.

Vol. 43: B.C. Arnold, Majorization and the Lorenz Order: A Brief Introduction. vi, 122 pages, 1987.

Vol. 44: D.L. McLeish, Christopher G. Small, The Theory and Applications of Statistical Inference Functions. vi, 124 pages, 1987.

Vol. 45: J.K. Ghosh (Editor), Statistical Information and Likelihood. 384 pages, 1988.

Vol. 46: H.-G. Müller, Nonparametric Regression Analysis of Longitudinal Data. vi, 199 pages, 1988.

Vol. 47: A.J. Getson, F.C. Hsuan, {2}-Inverses and Their Statistical Application. viii, 110 pages, 1988.

Vol. 48: G.L. Bretthorst, Bayesian Spectrum Analysis and Parameter Estimation. xii, 209 pages, 1988.

Vol. 49: S.L. Lauritzen, Extremal Families and Systems of Sufficient Statistics. xv, 268 pages, 1988.

Vol. 50: O.E. Barndorff-Nielsen, Parametric Statistical Models and Likelihood. vii, 276 pages, 1988.

Vol. 51: J. Hüsler, R.-D. Reiss (Editors). Extreme Value Theory, Proceedings, 1987. x, 279 pages, 1989.

Vol. 52: P.K. Goel, T. Ramalingam, The Matching Methodology: Some Statistical Properties. viii, 152 pages, 1989.

Vol. 53: B.C. Arnold, N. Balakrishnan, Relations, Bounds and Approximations for Order Statistics. ix, 173 pages, 1989.

Vol. 54: K.R. Shah, B.K. Sinha, Theory of Optimal Designs. viii, 171 pages, 1989.

Vol. 55: L. McDonald, B. Manly, J. Lockwood, J. Logan (Editors), Estimation and Analysis of Insect Populations. Proceedings, 1988. xiv, 492 pages, 1989.

Vol. 56: J.K. Lindsey, The Analysis of Categorical Data Using GLIM. v, 168 pages, 1989.

Vol. 57: A. Decarli, B.J. Francis, R. Gilchrist, G.U.H. Seeber (Editors), Statistical Modelling. Proceedings, 1989. ix, 343 pages, 1989.

Vol. 58: O.E. Barndorff-Nielsen, P. Blæsild, P.S. Eriksen, Decomposition and Invariance of Measures, and Statistical Transformation Models. v, 147 pages, 1989.

Vol. 59: S. Gupta, R. Mukerjee, A Calculus for Factorial Arrangements. vi, 126 pages, 1989.

Vol. 60: L. Györfi, W. Härdle, P. Sarda, Ph. Vieu, Nonparametric Curve Estimation from Time Series. viii, 153 pages, 1989.

Vol. 61: J. Breckling, The Analysis of Directional Time Series: Applications to Wind Speed and Direction. viii, 238 pages, 1989.

Vol. 62: J.C. Akkerboom, Testing Problems with Linear or Angular Inequality Constraints. xii, 291 pages, 1990.

Vol. 63: J. Pfanzagl, Estimation in Semiparametric Models: Some Recent Developments. iii, 112 pages, 1990.

Vol. 64: S. Gabler, Minimax Solutions in Sampling from Finite Populations. v, 132 pages, 1990.

Vol. 65: A. Janssen, D.M. Mason, Non-Standard Rank Tests. vi, 252 pages, 1990.

Vol 66: T. Wright, Exact Confidence Bounds when Sampling from Small Finite Universes. xvi, 431 pages, 1991.

Vol. 67: M.A. Tanner, Tools for Statistical Inference: Observed Data and Data Augmentation Methods. vi, 110 pages, 1991.